Social Issues in Technology:
A Format for Investigation

SOCIAL ISSUES IN TECHNOLOGY
A Format
for Investigation

PAUL A. ALCORN
Associate Professor
DeVry Institute of Technology
Atlanta, Georgia

Prentice-Hall, Inc., Englewood Cliffs, New Jersey 07632

Library of Congress Cataloging in Publication Data

Alcorn, Paul A.
    Social issues in technology.

    Bibliography: p.
    1. Technology—Social aspects. I. Title.
T14.5.A43 1986   306'.46   85-6517
ISBN 0-13-815929-7

Editorial/production supervision: **Diana Drew**
Interior design: **Jayne Conte**
Cover design: **Ben Santora**
Manufacturing buyer: **Gordon Osbourne**

T
14.5
. A43
1986

Printed in the United States of America

10 9 8 7 6 5 4 3 2

ISBN 0-13-815929-7 01

Prentice-Hall International (UK) Limited, *London*
Prentice-Hall of Australia Pty. Limited, *Sydney*
Editora Prentice-Hall do Brasil, Ltda., *Rio de Janeiro*
Prentice-Hall Canada Inc., *Toronto*
Prentice-Hall Hispanoamericana, S.A., *Mexico*
Prentice-Hall of India Private Limited, *New Delhi*
Prentice-Hall of Japan, Inc., *Tokyo*
Prentice-Hall of Southeast Asia Pte. Ltd., *Singapore*
Whitehall Books Limited, *Wellington, New Zealand*

To **Martel** and **Cindy Day**
with my thanks and my love
because dreamers are the saviors of the world

Contents

Preface

The subject of technology and the ways in which it interfaces with society is a vast body of knowledge. To adequately cover it would require many volumes of historical and contemporary information. It is because of the vastness of the subject that this book has been written. The purpose here is not to offer the reader a survey of past, present, and future technological innovations and their effects on the world, but rather to inspire the readers to study those technologies for themselves, learning by doing, understanding by experiencing, and developing the expertise necessary to predict and prepare for the future.

This book was originally produced as an introductory text for a seminar course in the field of social issues in technology, presenting a format of study through which students could investigate topics for discussion and then relate those topics to the general social structure.

Beyond this purpose, this text is a blueprint for survival for anyone who wishes to be the master of his or her own future. Written in a simple and straightforward manner, the book challenges those who realize the importance of the increasing proliferation of technological changes in the world today, and who understand how the changes promise to affect their lives, to begin thinking in terms of the future. Tomorrow is where we will all spend the remainder of our lives, and tomorrow is the home of technology. The rate of change that exists in our society today is by no small measure the result of our ability to manipulate our world to suit our own purposes. This "technologizing" is the means by which we mold our society and define the relationships that will or will not exist there. The need to understand those changes before they occur requires a shift in context in the way we view our world. The purpose of this book is to provide that contextual shift, to create in the mind of the reader the image of tomorrow's world, and to encourage the reader to think in terms of how that world will be different from the one we now live in. Only in this way can we be prepared for our own futures. Only in this way can we remain in control of our destiny, prepared for change before it occurs, armed with an understanding of what is taking place and how we choose to react to the changes that occur.

The material in the text is presented in three sections. The first deals with some underlying principles of the creation of technology and how it relates to the culture. The emphasis in this section is on the nature of technology, not specific technologies.

In the second section of the book, methods by which readers may launch into their own studies of the subject are offered, including a general view of the cause-and-effect, models, simulations, and systems approaches.

In the third and final section, a checklist of social factors is presented, indicating their relationship to technology. The unity of society and technology is also discussed.

Throughout the book, experiential exercises are offered to aid the reader in practicing what is presented in the chapters. These are not exercises with definitive right and wrong answers, but rather educational experiences to help the student understand the processes of creating and maintaining technological environments. The goal is to teach the students how to teach themselves. The purpose is to create an opportunity to explore what technology is all about in a societal context and to gain an understanding concerning its mechanizations as it helps create the universe in which we live.

By way of acknowledgments, the author wishes to thank the staff and editors of Prentice-Hall, without whose vision this work would never have been possible; my colleagues, for their understanding and willingness to share their views; and especially my wife, Aileen, who, as always, has ever been the patient and enthusiastic helpmate through the months of study and toil.

Paul A. Alcorn

Social Issues in Technology:
A Format for Investigation

PART I

The Nature of Technology: Foundations

CHAPTER 1

Technology: A Natural Process

He was a small creature, no more than forty inches in height, and he was, as usual, hungry. His brain was about half the size of a modern Homo sapiens, about 650 cubic centimeters, and he lived on a wide, flat plain in Africa, where the tall grass and clutches of low-branched trees made a hunter's paradise for him and his fellow creatures. His kind would one day be known as Homo habilis, but that was nearly two million years in the future. For our purposes, we will call him George.

As I have said, he was hungry. It was a characteristic of his species. Warm-blooded and a hunter, George and his companions spent much of their time ranging out across the plain near their most recent campsite in search of food. They were omnivorous, as likely to devour the tough nutty fruit of a nearby berry bush as the raw flesh of some small reptile or insect that failed to escape their notice in time. From day to day, George and his fellows satisfied their internal furnace with the fuel of whatever they could find, always searching for the great kill that would allow them to gorge themselves and replenish the dwindling supply of protein gathered from the last great kill some days or even weeks earlier.

Today they were near the high rock carapace to the east, though they had no concept of direction in those terms. It was merely the "high place over there, where the sun rises." George was scouting ahead of the pack, a chore he seemed to relish. A loner, he would often run ahead, somehow enjoying the prospect of being the first to sight a potential prey, hoping to be the first to wrestle it to the ground, to pound it to death with his clenched hands, or to tear its throat with his teeth.

The band was in the narrow passage that led into the center of the mound of rocks, close to where a fellow hunter had perished only days before at the hands of another predator, a huge cat creature with claws to tear at the throat and jaws to sink deeply into the flesh and break the victim's neck. George had seen it happen.

He remembered it all too well. He was cautious, listening and sniffing the air, remembering what had happened to . . . who was it? His simple consciousness forgot those things easily, but the memory of the danger remained solidly in his mind.

The others were far behind him and out of sight as he turned into the natural bowl formed by the circle of high, flat rocks near the center of the carapace. He could feel the eyes on him, almost smell death in the air. Instinctively, he knew he was not alone. Turning quickly, he scanned the rocks above, seeking any telltale clue of whatever was lurking there. He spun so quickly and jerked his head about so violently in his panic that he nearly missed the low, flat, black furred head, the huge yellowish eyes that stared back at him.

Above George and a little to his left was the same sleek creature that had made a meal of his fellow hunter only days before. George panicked. He turned and leaped to the side of a sheer rock, clinging with his toes and fingertips as the cat made a lunging pass at him. The panther missed the small ape-man by inches. George scrambled toward the summit of the rock, churning his legs wildly in search of some foothold, reaching out blindly with his hands for any purchase further up the rock face. Lacerations appeared on his knees and thighs as he slid against the sharp black obsidian. His fingers numbed as they bit again and again into the narrow, knifelike crevices above. But he was making progress. Below him, the cat yowled and paced, panting heavily and leaping toward the fleeing figure.

Springing with all its might, the huge panther nearly reached George, who pulled forward with a last great effort and reached a wide ledge nearly half way up the rock face. As he slid himself onto the strip of rock and flattened himself against the wall, a single stone slipped over the edge and fell, striking the huge cat squarely on the nose. With a howl, the panther retreated. George kicked another rock toward the beast, then another, missing both times. In panic, he grabbed several more and hurled them toward the beast, striking him again, this time dead center at the skull. The panther slumped to the ground, stunned by the blow. George grabbed for another loose rock and another, improving his aim with each throw, grasping larger and larger rocks until at last he found himself holding heavy slabs of obsidian over his head with both hands and hurling them down on the lifeless victim. He struck the creature again and again and again.

In the night, belly taut, legs splayed out before him, George lay with the other hunters in the natural bowl of the rocks, stuffed with the meat of the dead panther. He was smiling, staring up into the night sky at the bright starlit veil, a swath of white that spread across the sky like a river in the firmament. Absently, he licked his hand and passed the thick saliva over the crusted scratches on his belly and legs. Around him, the sound of night creatures echoed off the surrounding walls as predator and prey continued the struggle for survival.

In his right hand, George clutched at a single round stone, about three inches long across its short axis and weighing nearly half a pound. He felt safe now. He knew that he could fend off any attack. Tomorrow he would try his luck again with his newfound weapon. Tomorrow he would try it against one of the doglike scavengers of the plain or use it to bring down a bird near the river. Perhaps he would never need to be hungry again. Had he not slain the mighty panther single-handedly? Had he not proven himself the greatest hunter of them all? Who knew what he might be able to do the next time? Who could really know?

Was this the beginnings of technology? Probably not. Can we infer that George is the precursor of all humanity's inventiveness? This is definitely an erroneous assumption. Is this then only a story? No, it is a great deal more than that. It is a concept. It is an admittedly fanciful presentation of a process, a process that has inexorably led to the technology we enjoy today.

In the imaginary actions of this one Homo habilis, a creature who populated the broad fertile plains of Africa more than two million years ago, is the seed of all that we are as a civilization today. What happened in this brief incident is what happened again and again in our own real past. It was a first step in a chain of events leading from complication to complication, from one level of sophistication to new levels of sophistication. Now, in the twilight of the twentieth century, we find ourselves on the edge of ultimate power over nature—the power to destroy or enhance not only our own lives but the very existence of every living creature on the planet. This is due in large part to what we call *technology*.

The question we need to ask first, the one that must be explored before we can ever hope to understand this process which has so totally enmeshed itself in every aspect of our lives, is simply whether or not technology and all that comes with it is a natural consequence of our humanity or an artificial construct, separate from the natural way of doing things, and, perhaps, in direct opposition to nature itself. Are we dealing with an attempt to thwart nature, or are we expressing a part of what we naturally are? What of steel mills and electrical power, computers and steam engines, printing presses and gears, valves and mortar, and chisels and plows? Are all of these artifacts of our technological development unnatural? Are we dealing with man fighting against nature? Or are we merely seeing a *natural progression* of the species along an evolutionary path?

It is, as one television advertisement would have us believe, "Not nice to fool mother nature." Fighting the entire universe, it would seem, is a hopeless task, at best an irrational idea. Is that what we are doing? This is the issue to which we will initially address ourselves.

THE EVOLUTIONARY PROCESS

In the description of our fortunate Homo habilis, George was able to grasp the meaning of a combination of events and use them to advantage in preserving his life. The fact that he was capable of doing that stems from the fact that nature had supplied him with a brain capable of making those vital connections. Without this ability, he would most likely have perished. The facility to reason, which is what we are really dealing with here, is a survival trait, that is, a characteristic of the organism that enhances the creature's chances of survival in a hostile environment. This is part of what nature does, and it is part of what has come to be known as evolutionary theory.

Evolutionary theory has become a hotly debated subject in recent times. Whether its tenets explain the development of humankind accurately is not always clear, particularly in the eyes of those who champion alternative theories. But it is a useful theory, one that tends to support the bulk of scientific evidence and one that is increasing in sophistication and refinement. It is not the purpose of this book to attack or defend the evolutionary theory, nor to argue the vagaries of that part of its content beyond the periphery of our understanding. The theory has proven itself to be an excellent method of explaining real-world phenomena and as such is here accepted in a simplified form as correct. Of particular interest to us in our study of technology and society are the concepts of "survival of the fittest," "natural selection," "specialization," and "adaptability."

Natural selection and survival of the fittest are passive phenomena, representing observations of the way in which nature appears to operate. Basically, these ideas say that there are a very large number (if not infinite) of ways in which organisms can exist but that only those best suited to surviving in an environment will tend to perpetuate themselves. Consider the many differences between a one-celled creature such as the amoeba and the complicated network of cells that go into the construction of a human being. Natural selection says that any other combination or any other path along a developmental scale is possible, but that this one exists because of its superiority as a "format" that works. Humans could have had gills. We could have had webbed feet and five-foot-long necks and compound eyes and exoskeletons and a host of other features. But we do not. And the reason we do not is that they don't work as well as what we *do* have. The other features, the ones we do not have, simply do not exist because they perform no useful function for us. We do not *need* gills or webbed feet or compound eyes, and therefore we do not have them. For fish, gills are fine. For insects, compound eyes work beautifully. For a giraffe, a long neck is essential for survival. But for Homo sapiens, they are superfluous.

Natural selection says that as traits occur, if they are useful, the organism with the trait has an "edge" over similar organisms without the trait and, through time, more of the organisms *with* the trait survive than those without (survival of the fittest). The result is a predominance of organisms with the trait and the trait becomes "generalized" over the species.

For purposes of illustration, let's use an analogous example from our own culture. In a given population, there are a wide range of physical and psychological types. Some people are physically strong and others physically weak. Some are extremely intelligent and others not so intelligent. Some people are brilliant when it comes to music, whereas others, who cannot carry a tune in a bucket, have the capacity to understand complicated theoretical physical laws. (That is not to indicate that the two traits are mutually exclusive.) Now let's take a random sample of these people and put them into a specific environment, for instance, an accounting firm. For the sake of discussion, a representative sample of people all go to work for the same accounting firm on the same day. Ignore for the moment that the firm itself is selective.

Certain facts become rapidly apparent. People who are physically strong but not particularly capable when it comes to mathematics will find themselves at a distinct disadvantage. Those who are physically small or who have only normal physical strength find the lack of superiority in this trait of no consequence when it comes to doing their work. Those who are blind find that they cannot probably function at all unless artifically aided in some way. Musical geniuses, unless they can find some way to use their musical skills constructively for accounting, are going to find that the additional trait, while admirable, is of little or no use to them in their job. At the end of a year, assuming that all of these people are competing for a limited number of jobs within the firm, the only people left at the accounting firm from the original sample will be those whose traits lend themselves to the job at hand, that is, accounting.

We could have as easily made a case for fifty people going out for a football team or fifty people entering college or fifty people aspiring to be mystery writers. Those who survive in these "environments" are those who are best suited to carry out the activities connected with the environment.

This is conceptually no different from what takes place in nature. The world is filled with creatures who compete for survival in an environment that is filled with the possibility of failure and destruction. Only those whose traits are such that they have an edge over their fellow contestants can survive.

The concept of specialization carries this process a step further. Given that those best able to survive in an environment are the ones who will survive in an environment, we find that nature will select more and more succinctly for traits that fit in with the environment in question. This is called *specialization* and it is the tendency of an organism, through time, to only retain those traits that lend themselves to survival and to give up or forego those traits that are not useful to this end. Economy is a basic law in nature, or so it appears. Waste is not a thing that nature tends to support. Everything is done in such a way that efficiency is maximized. It is one of the reasons that technology mimics nature so closely in principle. Nature seems to know how to get the most out of a system. In terms of specialization, this means that traits that are useful remain through time and those that are not fall by the wayside, thus maintaining the efficiency of the organism.

But specialization is a double-edged sword. On the one hand, it increases the ability of an organism to survive in a given environment by selecting those traits that ensure success. This is its positive side. On the other hand, by removing through the process of specialization those traits that do not lend themselves to survival in a specific environment, it makes the organism dependent on the environment itself for its survival. In other words, *as an organism becomes more specialized and therefore more efficient at survival within a given environment, it also becomes more dependent on the specific environment for its survival.* That means that outside the specific environment, it has difficulty surviving.

Let's return to our analogy of the accounting firm. We have seen that within the firm, the employees remaining after a year are the ones best suited to the environment, that is, those who have traits that match the needs of the firm. A good accountant will still be employed, whereas one who is only marginally successful will have fallen by the wayside. Now assume that the environment has changed. Suppose the accounting firm is absorbed by another, larger firm that replaces ninety percent of the jobs with computer technology and transfers most of the staff to another sector of the parent company specializing in forecasting and predicting future markets. What happens to the employees of the original accounting firm? Obviously, some will survive. Those whose traits are equally useful in fields of accounting and in forecasting will feel right at home. Some of them may even advance more rapidly in their jobs because of some unused traits that now come into play, such as intuition and the ability to think both inductively and deductively. Others, however, will find that some traits that they do not have are essential to survival in the new job, and they will eventually leave the company, either voluntarily or through being fired. If a person is a whiz with numbers and details, for instance, but has very poor social skills, he or she may not survive in the firm due to the inability to express his or her opinions in meetings, interact with other departments, or deal with the public. As long as the individual was insulated from these activities, he or she thrived. Faced with new responsibilities and requirements, however, he or she is forced out of the organization.

Back to nature. The same phenomenon occurs when a natural environment changes. Specificity in an organism created through the process of natural selection ties an animal tightly to its environment. If the environment changes, the results may be disastrous. The key to long-term survival appears to be the ability to change with the times, to adapt over time to a changing environment. If the environment changes too rapidly, or if the organism is too tightly specialized, it may become extinct before it can adjust. This explains the importance of the last concept of the evolutionary theory that we need concern ourselves with—*adaptability*.

At this point, it is useful to note a concept that will be dealt with throughout the text. It is the concept of *balance*. From what we have seen so far, two things are necessary in order for an organism to survive in nature—the ability to fit into the structure of an environment through natural selection and specialization and the ability to adapt to changes in the environment. These two may appear to be somewhat in opposition, specialization defining a very tight system and adaptability describing the necessity of the ability to despecialize. In truth, what is necessary for survival is the proper combination of specialization and ability to change. Change implies the passage of time. Inasmuch as the environmental changes that occur are not too catastrophic or too revolutionary in character, an organism, though tightly specialized, may still adapt to the change through natural selection. It is a balanced combination of the two that allows a species to enjoy long-term survival.

Nature is always recreating a state of balance, or equilibrium, as it functions. Evolution is itself a function of equilibrium, constantly redefining life in such a way that it can harmoniously fit into the fabric of the biosphere. When change occurs, nature adjusts, and change is, after all, the one thing we can be certain will take place. Organisms change. Others cease to exist. New life forms come from old forms to thrive side by side with successful organisms already extant.

The world is full of examples of animals and human cultures that have adapted to their environment. This, at least, is fundamentally undeniable. Insects exist in forms that fit with the advantages and shortcomings of their environments. Animals live where they are capable of surviving, having those traits that lend themselves to the animals' survival and void of those traits that would endanger their existence. One does not find Gila monsters in Alaska, nor does one encounter polar bears in Arkansas. The flora and fauna of a given environment are "selective" of what works.

There is a beetle in Africa that enjoys living on termites. That in itself would not be particularly amazing were it not for the manner in which the beetle has become specialized to the task. This particular beetle uses two methods of hunting that allows it to be extremely effective. One is its ability to disguise itself as a rock, and the other is its ability to use natural tools to enhance its prospects of success.

　　　　The beetle coats its shell with a sticky substance which it produces internally and then covers itself with small grains of sand and other debris until it resembles a pebble. It then positions itself near the entrance of the nearest termite nest and waits. Soon, worker termites near the entrance who, as part of their natural programming (what we call *instinct*), have a desire to keep the entrance to the nest clear, venture out to remove the pebble from the mouth of the opening. They unsuspectingly grasp the "pebble" and carry it to one side. At a distance sufficient to avoid notice, the beetle turns on the worker and kills it. And so we see the advantages of looking like a rock.

　　　　But our beetle doesn't stop there. Rather than devouring the freshly killed prey, it returns to the nest with the corpse and waves it over the entrance, attracting the attention of soldier termites, whose job it is to protect the integrity of the nest, enticing them out into the open. In this way, the beetle is able to supply itself with a wealth of victims, each coming willingly to the slaughter. Is this a natural process, or is it technology? Is it both?

　　　　Nature is filled with examples of creatures like the beetle that are able to adapt to their surroundings. Prairie dogs utilize natural law to create ventilation in their highly sophisticated towns. Beavers use engineering techniques as complicated as those discovered by humans to control the flow of water in rivers and streams. Birds build nests that depend on static tension and precise structural design for stability. Fish use lures ranging from phosphorescent lights to tongues that mimic worms in their search for food. Insects use camouflage for protection as do lizards and mammals. Ants and bees and baboons and Homo sapiens use complicated social structures for mutual benefit and defense. Why?

　　　　In every case, the answer is the same. They do it because it works! That is the bottom line. That is the value of nature's pragmatic approach to the evolution of the various forms of life that inhabit this planet, and it is as true of humankind as it is of any other creature on the face of the earth. Nature simply does what works.

GENETICS AND EVOLUTION

And how does nature accomplish all this restructuring of organisms and ecosystems? For the most part, it does it quite effectively through a genetic device, RNA. RNA is the encoder, the messenger of an organism that transmits information from generation to generation in order to assure that each generation knows what it is supposed to look like, how it is supposed to behave, and what is expected of its constituent parts. It is in effect a blueprint of the organism in question. All that an organism is, including the changes that take place within that particular organism, and all that it is capable of being are encoded in the genes of the RNA present in every cell of the organism. As differences occur in a single organism's makeup, that which makes it different from others of its own kind, the differences are recorded in the RNA codes, ready for transmission to the next generation issuing from that organism. If a given trait adds to an organism's survival ability, there is a greater probability that the particular organism will survive long enough to reproduce itself, including the new characteristic. This is nature's method of transferring instinct from one generation to the next. Behavior patterns that are instinctual are transmitted to each new generation of an organism through the functioning of this encoding system. It is, as we have seen, extremely effective. Every living thing is capable of being and doing what is successful (what works) because of the instinctual programming transmitted through this process. Thus a duck flies south for the winter due to an instinctual understanding that that is what it is supposed to do. No one needs to teach a moth how to weave a cocoon, or an angler fish how to dangle its "bait" in front of its mouth to attract a potential victim. No one needs to explain to a spider monkey that remaining in the trees reduces the number of predators it need concern itself with, or to tell a spider how to weave a web in the most efficient fashion. These things are understood instinctually. This "biological programming" gives creatures an edge within their own realms.

And yet, instinct is restricting. If instinctual programming is thwarted, if it is somehow blocked, the creature cannot, in many instances, alter its behavior. The results may be disastrous.

Consider the sand wasp. Nature has supplied this insect with several instinctual behavior patterns that allow it to be highly successful as a species. The sand wasp digs its nest in the ground in, as its name indicates, loose, sandy soil. The nest itself is little more than a small hole in the surrounding landscape. It is advantageous for the sand wasp not to advertise the location of its nest. Predators can enter and devour its young. Others may take over the nest, awaiting the wasp's return to make a meal of it. It is not easily seen (as anyone who has inadvertently stepped on one when the resident was home can tell you).

Yet sand wasps always know where their nests are when returning from foraging expeditions in the surrounding territory. How can they be so sure? Research indicates that the sand wasp fixes the image of the nest in its mind by recording its relationship to surrounding landmarks. It has a record of the nest's location firmly fixed in its mind and can always return to the same spot after an expedition. This is confirmed by covering the hole. Even when a secondary false hole is added, the sand wasp will go directly to its real nest and uncover it, ignoring the false nest entrance. But if the landmarks around the nest are moved, if pebbles and plants are rearranged, the sand wasp can become confused, be unable to find its nest again, and become totally undone by the fact that the landmarks are not where they are supposed to be. It may never find its nest again. Because the method of locating its nest is instinctual, *it has no choice as to how it goes about the business of locating it.* And therein lies the key to the difference between humankind and other organisms.

Evolution and Artifact

So far, we have taken a brief look at some of the basic concepts of evolution. We have explored in a general way how evolutionary theory explains the diversity of life (humankind included), that exists on the planet. We have looked at the importance of specialization and natural selection to survival and have seen how instinct can aid a creature's survival capabilities. Yet we have said nothing of technology except by way of analogy. Or have we?

What about the use of a dead termite to lure soldier termites from a nest? Is the beetle not using a tool in its search for food? Is this an artifact, that is, an "unnatural" device? Is this a primitive example of manipulation or is it an adjustment to an environment? A beaver fells trees and builds elaborate dams to divert the course of a stream and create a home for itself and its mate. Is this the utilization of engineering principles to achieve an end? Is technology at odds with instinct?

Obviously, the beetle using a dead termite as a lure and the beaver building a dam are behaving instinctually. Neither of them have been to college to learn how to do these things, nor have they sat at the knee of an elder around a warm fire, learning what works. So what has all this to do with technology and the way in which humans operate in nature?

The difference lies in *the way humankind goes about the evolutionary process.* The transmission of information concerning what is successful and unsuccessful in a given environment is done in most cases through the genetic process, through RNA coding. We call this *instinct.* When adaptation is necessary, the vast majority of creatures adapt slowly, over many generations, allowing for a natural selection of positive survival traits. Human beings, on the other hand, have been given a survival trait that allows them to bypass this procedure and adapt with a speed that is incomparable when compared with other animals. Turtles have hard shells because shells help them survive. The shell protects them from predators. Birds have feathers and fly through the air because this gives them an advantage over earth-bound creatures. In the case of the turtle, we see a trait so successful that it has remained for hundreds of millions of years. In the case of the bird, we see an adaptation that took millions of years to create.

Birds originated with a small coelurosaur known as an archaeopteryx, a theropod living during the late Triassic and early Jurassic Periods. The small reptile, finding itself in competition with the emerging mammals, many of whom were nocturnal predators, developed feathery scales that eventually developed into feathers, allowing it to survive in the face of increased competition by escaping into the skies. It took nearly thirty-five million years to accomplish this! Humanity accomplished the same feat without growing the first feather.

The difference is that the human being evolved externally to his/her body as well as internally through the process of natural selection. Marshall McCluhan expressed it succinctly when he referred to wheels as extensions of the feet, cameras as extensions of the eyes, and microphones as extensions of the ears. It is this external evolution to which I am referring. Humanity is specialized in one respect—the ability to think abstractly, which allows the species to be at once specific and general to an environment. We do not have long, sharp teeth as do members of the cat family, yet we are much more efficient hunters. We cannot run at more than sixty miles per hour like the cheetah, yet we can move much more rapidly and for much longer periods of time. We are not the biggest, nor the strongest, nor the swiftest. We are not equipped with natural camouflage or repellent chemical sprays or hard-shelled armor. We have none of these specializations, yet we are one of the most successful creatures in nature. And that is because we are adaptive.

Rather than grow evolutionary changes, humanity manufactures them. In order to adapt, we *extend* ourselves out into the surrounding environment, not being restricted by internal change. With the spear as a weapon, we need not have sharp, extensive canines. With the bow and arrow and later the flintlock and rifle, we can extend our "fangs" far beyond the range of any other creature. With the wheel and steam and the internal combustion engine, we expand our mobility at a rate that surpasses any other beast on the planet. In each case, we are extending ourselves outward from our bodies into our environment, at the same time changing the environment within which we are operating in order to suit our own purposes. If it is too hot, we create refrigeration. If it is too cold, we learn to control fire, to build the hearth and chimney. Each extension increases our ability to adapt to change and to react to alterations in the environment that we may encounter. In effect, we have specialized in the ability to generalize.

Humankind does evolve. We do adapt to changing environments and changing conditions. But it is accomplished through the creation of artifacts, by a manipulation of the environment and an understanding of natural law that enables us to improve and recreate the environment in the form we find most advantageous.

The archaeopteryx developed from dinosaur to bird in thirty-five million years. Human beings progressed from Kitty Hawk to the moon in less than fifty. As an evolving species, we seem to always be in a hurry. We seem to be too interested in progressing to wait for nature to supply us with the internal devices that we need to progress. We do not need specialization in the physical sense because we are specialized in one very important way—the ability to think. Conceptually, there is no difference between a turtle's shell and a human's "house," although the latter is more flexible in form and function. It is still a protective dwelling. The concept is the same although the application is greatly expanded in the human body. Conceptually, there is no difference between a bird's wings and an airplane. Both are means of flight. Yet the variety of designs and uses to which humankind has put the concept of flight is far beyond that of birds. We evolve rapidly and we evolve externally. We use artifacts, that is, artificial constructs that mimic natural principles, to do the job rather than physical changes. And because of that one fundamental difference, we have been able to conquer every environment and every circumstance the planet can offer, from the depths of the ocean to the cold of the poles. We generalize by being specific only in our ability to understand, to construct technology, and to adapt. And the evidence of how successful this approach can be is made clear by the huge numbers of human beings and their influence over the world's ecosystem.

CONCLUSION

What has been shown in this chapter is that humanity is as much a part of nature as any other species, and that this includes its use of technology and artifacts and its ability to manipulate the environment to suit its own needs. Mankind's actions are often involved with the production of "unnatural" devices that we call *artifacts*. They follow natural laws and perform useful work in accordance with the way the physical world is structured. They are artificial by definition in that they do not appear naturally in nature. This does not mean that technology is not a natural part of the evolutionary process. Nor does it mean that technology represents an unnatural refusal to live within the ecosystem or that what we do is in opposition to the equilibrium or balance of nature. We do what every other organism does: We evolve, we adjust, we change, we react. It is only in the *external* manner of our evolution and in the far greater efficiency of adaptation that we are separated from other animal life. Technology is part of what it is to be human and, consequently, part of what it is to be part of nature. We are an "experiment." We are an attempt on the part of nature to look for a better, more efficient way of doing things. We are a step on the road to order. So far, we have been a very successful experiment. Whether or not we will continue to be so remains to be seen. This is a large part of the subject matter that makes up the rest of this book. We will start an investigation of this question with the next chapter, which deals with the "flip side" of the picture, nature's system of checks and balances and how the human race controls its evolutionary process.

SUMMARY

Each species in nature strives to match itself to the environment within which it operates in order to survive. It is a matter of matching skills with the biosphere, so that the particular specializations which a given species has selected are survival skills that will enhance its opportunities to reproduce and thrive. The measure of how successful a specific species is in this attempt is how long and how extensively it exists in the environment in question. For most species, this is accomplished through genetic structuring and programming and through the trial-and-error methods of natural selection. That is, those organisms that pick the proper set of characteristics and abilities will survive, while those that do not will fail to survive.

This is as true of Homo sapiens as it is of any other species. The difference between man and other animals lies in the way in which the organism goes about the processes of adaptation and specialization. Our specialization is not to specialize other than in the ability to think, to create artifacts, and to mimic nature to our own advantage. We are, by nature, generalists who adapt on a continuing basis to changes in the environment. We use the creation of artifacts, tools that are by their very nature not natural, to change rather than doing it internally, through a biological process. The result is rapid change, a rapid adaptability that has allowed the human species to fully populate the planet, and a dominance of the biosphere in which we are living. Animals with as wide a dispersion over the face of the planet as the human race tend to be dependent on human beings, that is, they adapt by becoming dependent on humans for survival, at least in part, as do the cockroach and the common rat. Both are found wherever human beings are found and in many cases directly *because* humans are there. The conclusion is that humanity is performing a natural function in "technologizing," which is a survival trait, even though the technological artifacts created in the process are not, in and of themselves, natural, being rather copies of nature and examples of the human race's ability to manipulate natural law to its advantage.

THOUGHT AND PROCESS

1. Primitive human beings experience their environment and learn from it, gaining the tools that will allow them to survive. What are some of the obvious connections between the probable observations of the primitive human being and the invention of (a) the sling, (b) the spear, (c) cooking, (d) the wheel, and (e) artificial shelter (houses)?

2. If invention and technologizing are environment-specific, speculate on the difference in primitive tools created by people living (a) in a tropical jungle, (b) on a heavily wooded tropical island, (c) on an arid, high-altitude plain, and (d) on a flat tundra near the Arctic Circle but not on a coast. How do your speculations compare with your knowledge of such cultures? Did you base your speculations on your knowledge of these cultures?

3. Suppose that you are marooned in an unfamiliar environment, for instance, having crash landed on an apparently uninhabited earthlike planet. You are surrounded by trees, wooded valleys, and a water source in the form of a small stream. You have no knowledge of the seasonal variations of the planet, what season you are in, how long you will be marooned, what the fauna are like, or if there are intelligent inhabitants on the planet. There are no hard minerals, but an abundance of sand, coal near the surface, and sandstone outcroppings. Describe your efforts to survive.

4. In light of the statement in this chapter indicating that adapting to a single environment dooms an animal to dependence on that environment, classify each of the following organisms in relation to their migratory capacities, arranging them in descending order of adaptability to a change in environment: (a) human beings, (b) elephants, (c) koala bears, (d) common honeybees, (e) mosquitoes, (f) grizzly bears, (g) bisons, (h) monarch butterflies, and (i) tree frogs.

   Now make another list based on sources of information on each of these creatures, and, in the light of your extended knowledge of their habits, compare the second list with your original.

5. If humans mimic nature in creating technology, we can assume that the microphone is a mechanical mouth to the speaker and ear to the listener, the wheel is an improved foot, the telescope and camera represent extensions of the eye, and so forth. How do we extend our concept of the human thinking process through artificial means? How do we artificially mimic the procreation–gestation–birth process? Are there analogies in human technologies that closely parallel these aspects of human biological life?

CHAPTER 2

Resistance to Change

William Fletcher was not an evil man. Quite the contrary, he was one of the most highly respected men in the district. He had grown up among the people in Harrogate, and was a part of everything that was typical of the traditional English countryside. Prior to this moment, he had never so much as considered an act that could be construed as unlawful, at least, nothing beyond the usual exuberance of his youth. Yet here he was, skulking about the commons like some miscreant up to mischief. He was beginning to regret his decision already. Harrogate was, after all, his home.

He loved this quiet village along the upper reaches of the Thames. The river coursed gently through this section of the country, moving so slowly in places that the current could scarcely be detected at all. But along its near shore, eddies and subtle changes in the bottom would churn it into a fearful rush, and it was that fact that brought William Fletcher to be abroad this night. It was also that fact that had brought the mill to Harrogate.

William had lived his entire life among the simple folk of Harrogate. Indeed, his family had lived and worked in this village for as long as anyone could remember. His very name bespoke his ancestors' honorable profession, first as arrow makers to a noble lord and later as armorers to the king. But that was all before. That was before the machines, before the new ways, before the coming of the landlord's new enterprise. It was before the factory had come to the banks of their river. Everything was different now, and William was not alone in his fear of the changes.

William had spent considerable time making his way around the commons, avoiding the more direct path across the green that would have meant detection by the squire's livery. He slipped from wall to wall, skirting stone cottages and deserted lanes until he found himself directly opposite the Hanged Horse Tavern. A final skittering dash to the rear of the building and he had arrived. Three sharp raps and he was through the door to Jonathon Roberts' private rooms.

William looked about. He was in a long, narrow room with a low, timbered ceiling and heavy planked floor. A single small window—all diamonds and circles where scraps of glass had been rescued from the bottoms of discarded bottles to give precious little light on a winter's day—punctuated the otherwise blank wall to his right. Candles stood in wooden holders by the entrance and by the door to the tavern itself. There was not a sound from the common room. No man would be about this late, at least, no honest man. The candles were unlit. The only light came from the burning log in the great hearth to his left. In its ruddy glow he could make out the silhouettes of nearly a dozen others, and he knew most of them well though he could not clearly make out their features.

Jonathon Roberts, the silent host, stood by the fire, his crooked leg as good a signature as any to distinguish him from the others. William still remembered the day the hay wain had fallen on Jonathon, crushing that leg. To Jonathon Roberts' left stood the Cooper brothers, Michael and Daniel, broad of shoulder and straight of limb, not yet bent by the work that would stoop their bodies in another ten years. To William's left was Allen Smithson, an iron monger like himself, and he had brought the three from Hardmoore as he promised. The remaining two were William's own brother, Harold, and his uncle, Geoffrey. William smiled and nodded.

"Are we all here, then?" asked Jonathon Roberts.

"All that be coming, I suspect," said Allen Smithson. "There'll not be many willing to do what must be done."

"Is it come to that, then?" asked William. "Is there nothing else for it?"

Jonathon Roberts heaved a great sigh and, as if in answer, the great log beside him on the hearth hissed and sizzled, sending a spray of red sparks into the air.

"You have seen as well as we, William Fletcher, what the squire's factory has meant to Harrogate. Nothing is the same. Nothing can ever be the same again if we do not act and soon."

William nodded assent. "Aye, Jonathon. We've said it all before. It is not a happy task we have before us now. I merely hoped . . ."

"That we could find some other way? Yer daft, man. Look about you. Our river's fouled with the waste of that evil enterprise. Aye, even down to Hardmoore, near seven miles away, it's fouled. We've suffered our peace disturbed, our lives disrupted, and our very livelihoods taken from us by those devil machines. There's no life to them, William. They're all wood and iron and stink and noise. There's no heart to them. The ways of our fathers have been destroyed. They're taking our young men to work and causing our women to yearn after rich-made goods. First they woo us with their promises of good wages and a better life, then, when new machines do our work faster, they send us on our way or lower our wages until we cannot support our families. There's not been an honest day's work done since their coming! We've not had a moment's peace since that day and we'll have none until we've destroyed them."

"But it's simply not right, Jonathon," protested William. "This is not our right! The factory and all within it belongs to the squire, aye, and our homes and livestock as well. This is not the way."

"Then why have ye come if ye feel that way?" retorted Jonathon.

William stood for a long moment in silence, looking into the dark face of

Jonathon Roberts. He could not hear a sound in the room but his own breathing and the crackle of the huge oak log in the fireplace. His shoulders finally slumped in resignation and he nodded, "Because I know of no other way to fight it, Jon. That is why I am here, as you well know."

"As I well know," Jonathon repeated. He crossed to the rough table in the center of the room and grasped the heavy iron bar laying there. He raised it into the air and brought it down on the table with great force. A plank split, sending a cloud of splinters and dust into the air. "This is the only way to stop them, William. This and fire. And once we've started, others will join us, of that you can be certain! We'll stamp out these factories here and now and have done with them! Let the progress be gone! Progress indeed! It's a lord's way of tramplin' his people underfoot! We'll send the squire's devil machines to hell where they belong!"

"And if the squire opposes us?" William challenged.

Roberts looked up into William's eyes with a deathly grin. "Then we'll send him to hell, too!"

Allen Smithson grabbed an iron bar from the table. The three men he had brought from Hardmoore lifted the axes that had been resting by their sides. William shuddered inside, closed the last door on his doubts, and licked his dry lips. He raised the heavy iron-headed hammer he had brought from his shop.

One by one, the men filed from the small room and began the short journey along the river chase to the factory. William wondered what the squire's men would do if they caught them. They were armed these days. They had been armed since the riots in Birmingham. He wondered if he would ever see his Emma again; if he would live to see tomorrow's first light. But then he remembered the gentle river and the quiet, unchanging village and the way his neighbors had strayed from their peaceful, ordered lives, and nothing else was of consequence. Only the factory and the machines mattered . . . and their destruction.

The scene depicted above has a strange flavor of both abhorrence and sympathy. On the one hand, we are presented with a picture of violence about to be unleashed on society, yet we may be strangely moved to sympathize with the plight of these unfortunate people, caught in the tangle of their dilemma.

This is a story of resistance, of frustration, of violence. It is a story of desperate individuals about to destroy technology. How can such a reaction come about? What forces people to take such risks? What compulsion could be so strong that they would go against all that they had been taught by the society in which they lived and commit crimes directly opposed to the strongest urgings of their ethical base?

The hypothetical workers depicted are typical of a group that actually existed in the early years of the nineteenth century in England. They were called *Luddites*, a name derived from the actions of one Ned Lud of Leicestershire, who, around 1799, destroyed two textile frames belonging to his employer. Ned Lud might be excused his excessive frustration over the introduction of labor-saving devices, but what of the Luddites to follow? Between 1811 and 1816, the Luddites routinely smashed labor-saving devices in the textile industry of England in protest of lower wages and unemployment stemming from the new technology's introduction.

The Luddites are an extreme example of a natural phenomenon that coexists with the development of technology, a phenomenon known as *homeostasis*. The word *homeostasis* comes from two roots, *homeo*, meaning the same or like, and *stasis*, which refers to the stoppage of the flow of change in some system. Hence, *homeostasis* refers to a resistance to change, a seeking of the status quo.

There are numerous examples of the homeostatic reactions of people to changes in society in general and in the technological base in particular. They appear throughout history and are probably far more common in our own lives than we suspect. Let's consider some other cases of homeostasis in action.

THE GREGORIAN CALENDAR

During the papal reign of Gregory XIII, a study revealed that the calendar system in use at the time, the Julian calendar, was inaccurate in its determination of dates. The inaccuracy resulted from the difference between solar time and the official calendar, a matter of one-fourth day per year. The difference had become so pronounced since the adoption of the Julian calendar in 46 B.C., that by the time Pope Gregory XIII looked into the matter, a total of fifteen days separated the calendar date from the true date. That is to say, it was fifteen days later than everyone thought it was. Much to Pope Gregory's credit, he developed our present calendar system, complete with leap years and other refinements, to take care of the disparity and to ensure that the lapse could not occur again. But what of the extra fifteen days? Gregory was a practical and well-organized man. He solved the problem logically and simply by starting the new calendar fifteen days forward of the old Julian date, thus eliminating the difference forever. So much for the problem of calendars. Or was it?

Consider how you would feel if you got out of bed tomorrow and were told by your favorite early morning news commentator that it was Monday, not Friday, because some famous scientists had just discovered after viewing the far reaches of the cosmos through their telescopes that the entire universe was being misinterpreted by a matter of four days. Accordingly, you had to go to work for five more days before the weekend rather than one. What would your reaction be? I don't blame you. I'd probably react the same way. And that's just how the common people of Europe reacted!

There were riots! There were burnings! There were *burgermeisters* and mayors put to the sword! And everywhere there was the cry of "Give us back our fifteen days!" And it was all because Pope Gregory wished to straighten out a long-time problem with the calendar.

COUNTERCULTURES

In the 1920s they were called *Bohemians*, in the 1950s we called them *beatniks* or the *beat generation*. In the 1960s they were called *hippies* and *yippies*, and in the 1980s, we call them *greens*. All of these labels have been used, either by those involved in the movements or by those opposed to them, to describe a phenomenon known as the *countercul-ture*. In each case, a group of people, usually predominantly young, chose to separate themselves from the mainstream of their society and try another approach, another way of doing things. They created their own social structures, their own literature, and their own dialects, seeking to disassociate themselves from the rest of the culture in which they live. And in each case, the roots of their "discontent" or desire to return to a simpler life could be traced to the technological upheavals of the times in which they lived.

Bohemians first appeared after World War I. They lived in a new world, totally different from the one that had existed prior to 1914, and, rather than move with the changing times, they chose to wander through Europe, living in the back streets of a city's old town and generally divorcing themselves from the progress going on around them. The "beat generation" sprang from the changes that took place in the United States after World War II, a period of redefining U.S. culture and lifestyle. The "beat generation" rejected the move to suburbia, the rise of the corporate world, and the move to higher and higher levels of affluence among the population epitomized by high-fidelity technology, television, and the suburb. They chose to spend their time philosophizing about the plight of the country from their dingy basement coffee houses, producing poetry and music that was marked, as much as anything else, by its alien themes and forms so different from the established patterns of the times. Then, in the 1960s, it was the hippies, with their long hair and their insistence on the simple life, free of encumbrances, a naive mixture of optimism and regression toward simpler times.

It is not the author's purpose to debate or pass judgment on the ideals and the trappings of these subcultures, each of which flourished for a time as a poignant statement of the troubles of our culture, but rather to view them as symptomatic of a phenomenon. It is the individual who refuses to accept the directions of society at large that is important to us here. The individuals in each counterculture had their reasons for their actions and for their attitudes. There are few people who can deny the effects of the hippie movement of the 1960s on our culture, both politically and practically. They were an instrumental part of the American reaction to the Southeast Asian war in Vietnam and neighboring countries. They undoubtedly contributed significantly to the ecology movement in the United States and elsewhere in the world. They enriched our art, our music, and our language, but above all, they demonstrated a fundamental refusal to follow the direction in which society was headed. In this sense, they demonstrated not only the truth of homeostasis, but also its importance to the process of technological change and our survival as a species.

Other examples of reaction to change could be cited, from the refusal of Russian peasants to accept agricultural improvements introduced by Peter the Great in the seventeenth century to the fears of technocracy in the 1920s. In each of these cases, we see the same demonstration of a resistance to a forward movement in the society punctuated by technological innovation. It is the causes of this homeostasis that need to be explained.

SOURCES OF TECHNOLOGICAL HOMEOSTASIS

The fear of technology and the accompanying homeostatic reaction are rooted in no single cause. A number of factors seem to contribute to the rising feeling of helplessness and fear that often accompanies advances in technology. The following discussion describes some major causes.

Fear of the Unknown

There is a basic experience that everyone who has ever lived has shared. It is the birth experience. And like all personal experiences, the cognitive processes of the mind allow us to assimilate information about the experience, using the knowledge to make determinations about our world. That is, when we experience something, we become aware of it at some level and place the new data gained into the overall fabric of our understanding of our universe. In short, we learn.

At birth, of course, we have no conscious understanding of what is taking place. We have no way to place the incoming data into perspective or compare it logically and classify it. This does not mean that we are not aware of what is going on, only that we have no way of communicating about it with ourselves or others, beyond bawling our eyes out, yelling our heads off, or cooing and laughing gently, depending on one's personal experience of birth.

There is an entire approach to giving birth that centers on the importance of the event as a psychological experience for the newborn and the parents. It focuses on the comfort of the child and on reducing the possibility of fear in the child to an absolute minimum. It is called the Leboyer method, and it is being used more and more in the birthing process.

Why bother? Why be concerned with the feelings of the newborn infant? After all, it has just been born and will have no memory of the experience, no recollection of who did what to whom. Why not just take care of the process as efficiently and effectively as possible, ensuring the well-being of mother and child? All babies cry when they're born, don't they?

As it turns out, this is not exactly true. The birth experience is far too often a traumatic one, and the importance of that fact is being increasingly recognized as a source of later behavior.

Consider this. You are an unborn child, safe in your mother's womb. Your entire experience is wrapped up in the gestation process during which you have nothing to do but grow and develop. In the next instant you are forced to emerge into an entirely new world, filled with unexpected and unfamiliar experiences. You are possibly cold; possibly in some pain; and constantly jostled about by unseen hands, the reality of which has no meaning for you. There are new sounds and new feelings far beyond anything you have experienced to that point in time. This is one of your very first learning experiences, and often it is a matter of learning that the unknown is frightening and therefore to be feared.

There you have it. Right out of the womb you are taught to be afraid of the unknown. From this birth experience, human beings often have their first taste of anxiety, and whenever new experiences come along, they may subconsciously return to that first experience of life that says that the unknown is to be feared. In this sense, homeostasis can be viewed as an attempt to avoid the unknown by avoiding change. No new experiences means no new bouts with the unknown. We can maintain our reference points. We know how to react and what to expect. We can avoid those pangs of anxiety that have been with us since very soon after birth. It is nothing more than the id in action, seeking pleasure and avoiding pain. What could be more natural?

If the birth experience were the only experience of change common to humans, however, they would simply never change. Fortunately, other changes occur in our lives before we are able to take any conscious control of our own circumstances, and through the learning process the young child develops a sense of understanding that all change is not bad. Indeed, there are many pleasurable new experiences as time goes on stemming from our interactions with our family and later with friends, and from our interactions with our environment through play. Play is, after all, a method of practicing for the real world to come. It's a training ground for how to handle the future with all its surprises and decisions. Thus we shuttle the birth experience into some neat corner of our mind and continue to impress new experiences and reach new conclusions about the meaning of their content, both good and bad, as we grow up. And this brings us to a second source of homeostasis, the process of *adjustment*.

Anxiety and Adjustment

Life is not always easy. And that is probably a very positive thing. Without resistance in our lives, we would have no reason to strive to overcome problems and there would be no growth, either in society or in the individual. In a hostile environment, the tendency to seek to overcome obstacles is a survival mechanism that works. It is the cause of technology in the first place.

But what about obstacles that cannot be overcome? What about when we are unable to overcome the difficulties inherent in some undertaking? Without some way to handle the possibility of our efforts being thwarted, we would be like wind-up toys, blocked in our progress by a wall or table leg, yet still pushing with all the strength of our spring-powered motors, trying to force our way through the difficulty until we run down or break.

Fortunately, most people neither run down nor break. Nature has supplied our wandering, creative minds with a series of defense mechanisms that protect us from overload. An entire branch of psychology, the psychology of adjustment, is devoted to the study of this one subject.

When we feel frustration we react by first exerting more effort toward the solution of the problem, causing more frustration. If this fails, we adjust in some way to the nonattainment of our goal. All of the various types of adjustment available to us are not equally acceptable in the eyes of society. This in no way prevents us from using them.

Resistance to change can be viewed as an adjustment process, by which the frustration and anxiety caused by the technological change are denied. People may feel helpless in the face of technology. It is new and it is undefined, an anxiety-causing condition that at first spurs them on to seek out knowledge that will alleviate their initial fears of the unknown. If and when this effort fails, then they must cope by adjusting to the change in some way in order to ease their feelings of helplessness. One of the forms that the adjustment can take is homeostasis. The Luddite-like villagers in the vignette at the beginning of the chapter were reacting in this manner. They were threatened and angered (another adjustment to anxiety) by the road progress was taking and, to combat it, they chose violent resistance. The countercultures of the twentieth century used withdrawal, a form of schizophrenic reaction, to deal with the changing environment within which they were living. They chose to withdraw not only from the technology but from the changing culture that the technology was creating. They "dropped out" and "turned on" as a means of escape from the realities of the changes that were occurring. This is not to say that all hippies were schizophrenic by any means. It merely indicates that withdrawal in the face of a technosociological change can be a defense mechanism designed to reduce anxiety and prevent the serious short-circuiting of a person's psyche.

Psychophysiological Restructuring

Still another source of resistance to changes in technology stems from the way we think physically, and the accompanying effects of this on the way we react. This has to do with the way we learn and the way we go about the process of relearning. It could best be summed up in the old saying about not being able to teach old dogs new tricks.

Actually, the saying is only half right. It *is* possible to teach old dogs new tricks; it just takes more time and effort to do so.

The human mind is a most incredible machine. Its efficiency in assimilating, manipulating, and handling information is astounding. In a newborn infant, the mass of the brain is present, ready to learn, ready to do its little tricks of reason and memory and manipulation, but the connections among the various brain cells through which these "tricks" are performed are undeveloped at best. The child has not as yet learned to *pattern* his or her brain to carry out higher-order functions.

The patterning, the arrangement of the networks of interconnections among brain cells, takes place as we learn. A newborn infant is indeed a blank page upon which the world will write. That writing or the impressing of information and thinking patterns occurs as we engage in interaction with our world. Many patterns are formed. Many more are possible. There is an extremely large number of networks that can be constructed among the billions of cells in the brain. The one that finally develops is dependent on the person's experiences in life.

Through the learning process, certain pathways are reinforced again and again, causing us to set up certain predominances in the way we think. If we deal with physical objects, such as balls and squares, blocks and long tubular beanbag worms, each time we experience these things our brain reinforces the pathways that were developed the last time we dealt with these same forms. This strengthening of specific connective networks gives us a certain view of the world that increases to predominance as it is reinforced over and over again. That is one of the reasons that memory is aided by repetition. Anything can be learned by the brain if it is presented in an understandable manner. By repetition, we strengthen the specific combinations of cellular pathways within the brain that are formed by the stimulating experience. It predominates. Something that happens again and again is far easier to learn than something that is only experienced once. To test that hypothesis all one need do is think of how easy it is to remember the letters of the alphabet (particularly if recited in order to the well-known accompanying tune) and how difficult it is to remember exactly what was happening in your life on this date four years ago at this hour of the day. Both thoughts involve remembering past experiences. The first is a repeated experience from early childhood, when the mind was plastic and just beginning to be written upon, and the second is not reinforced, and unless the circumstances of your life at that instant in time were particularly traumatic, it would be a monumental feat to recall what was happening.

As we grow older, we lose some of the plasticity of the mind, that is, some of its ability to rewire itself in the light of new information. As more and more experiences enter our life, we slip them into our consciousness (and possibly subconscious) through well-worn paths long used with success in running our lives. We compare and classify on the basis of our experiences. We fit new data into neatly ordered pigeon-holes in our minds.

What then of an alien concept that happens to come along? What if that concept is in opposition with our usual mode of thinking? What if it does not fit neatly into one of our little preconceived boxes? Are we able to understand it? Can we assimilate and place it into proper perspective? Psychologists tell us that we can, but that as we age, it becomes more difficult.

We are blessed with plasticity. We are capable throughout our lives of adjusting to new circumstances as they arrive. But the speed and ease involved in the adjustment, that is, the degree of plasticity that we as individuals have, tends to drop as we grow older. Reinforced concepts are more easily used than new ones. As we age, our learning curve takes a turn for the worse and we become more and more intractable when it comes to change and to new ideas.

Consider how traditional a typical seventy-year-old person can seem to be, how tenaciously he or she sometimes sticks to old ways of doing things, retaining old habits and old belief systems in the face of even the most incontrovertible evidence to the contrary. On the other hand, consider young children. They are ready to accept anything. They are curious, questioning, and willing to seek new knowledge wherever they can find it. The world is an adventure, a playground full of new ideas and new experiences. They seek it eagerly. They ask why and how and who about everything, and they can assimilate and use the new knowledge with incredible speed. It is because of their plasticity that this occurs. They have not had their experiences reinforced yet, as have their adult counterparts.

How does this apply to homeostasis? It means that new concepts, new ways of life, new technologies, and new social structures that accompany change are hard to accept when one is older. The mind is still plastic, to be sure, but it has trouble converting its belief systems. It is experiencing difficulty changing gears. And for that reason, much of the resistance to change that appears in the culture occurs in those who are older rather than younger. In order for a change to truly last in our world, it must fight and win over the objections of the adult population. They are the ones who will have the greatest trouble accepting it.

The computer and all it brings with it is the obvious example of this type of homeostatic reaction. A young child can learn BASIC in a very short period of time. In two or three sessions, the child can be merrily programming in a higher language such as BASIC, not at all put off by the strange jargonish nature of the commands or the strict step-by-step logic of the program's construction. Adults can and many do learn to use a computer as well, but the task is far harder. They have had no preparation in the sense that the child has. They have strongly reinforced patterns of neuronic interaction in their brain that determine the predominance of their mode of thinking. In many cases this is not a mode of thinking that easily lends itself to learning computer programming. It is hard work for the adult, not play. It requires effort, time, and concentration in the adult just to load and execute a canned program, constantly adjusting the adult's thinking to what the computer is doing. For the child, it is, what else, "child's play."

People resist that which is difficult for them. In our seeking of an easier way to do things, we tend to reject those ideas that are difficult in favor of those that are easy, and the familiar is always easier than the unfamiliar, whether it is the computer, a new form of music, or the workings of a mechanical loom. Frustration can result from the effort, followed by anxiety and finally a refusal to accept it.

This plasticity of brain networking also goes a long way to explain the predominance of youth in social movements. Society is a living organism, a symbiotic creature whose mechanism consists of all the people within it. It is only logical to conclude that the changes that the organism undergoes will be created by those best able to make the change. In the case of society, those are the young.

Chauvinistic Conditioning

Still another source of resistance to change is conditioning. It is the result of the plasticity problem discussed above, though it deals not with the inability of the individual to accept the change but rather with the preconceptions that a person may have as a result of the way in which that person views the change. We are dealing with a matter of context rather than content.

The *content* of a new technology is its physical reality. The content of steam power is the steam engine and the machinery it can run. The content of the computer consists of the software and hardware and the techniques for implementing them. The *context* of the technology is another matter altogether. This is the way one sees the technology based on preconceived ideas. Much resistance to technological innovation can stem from this contextual cause. Consider the person who suddenly realizes that his or her world is about to be peopled with robots. Robots will build the automobiles and operate the steel mills. Large robots will dig ore and run cranes. Small ones will serve drinks to guests in private homes or keep track of the physical environment within an office building. They may park your car, clean your park, and deliver your cleaning. There is no end to the labor-saving uses to which they may be put. And yet this person may be horrified, even if he or she fully understands the *technical* aspects of the new technology. Why? Because, as everyone who grew up in the 1950s and 1960s knows, *robots are monsters!* Conditioning. Reinforcement of ideas. Tales from childhood and motion pictures designed to make your hair stand up on end and science fiction novels created to fill your imagination with high adventure and great drama. And all of it ready to prove to you beyond a shadow of a doubt that a robot is a thing to be feared.

It does sound rather silly, doesn't it? It is. And it is also true. No matter how much you tell yourself that all that "stuff is bunk," you may still get a feeling of uneasiness when one of those mechanical monsters enters the room. If you have ever encountered a true robot under uncontrolled circumstances, you are probably already aware of this.

And the robot is only a single example. What about genetic engineering? Are we creating Frankenstein's monsters? What about the peaceful use of nuclear forces? Are we dooming ourselves to extinction? What about the image of technology as the destroyer of jobs? Are we putting ourselves out of business?

Much of the concerns we have about the burgeoning technologies of today are concerns born in ignorance and nurtured by the fears of others. Historically, technology has always resulted in progress. Historically, labor-saving devices both create and destroy jobs, though they create more jobs than they destroy. Historically, we have not created monsters through technological development unless, as with the tank and the nuclear warhead, a monster is what we were looking for. Yet conditioning may predispose us to a chauvinistic denial of reality in favor of our belief systems and thus cause an unreasonable resistance to change. It is the *unreasonable* nature of the fear born of ignorance that causes the resistance from this element of conditioning.

Specialization

Specialization is a fundamental underpinning of the Western way of life. It is a principle of economics and of life that has actually created much of the advancement that the human race has undergone in the past ten thousand years or so. It is a simple concept. Specialization means breaking the work up among a large number of people (what we call the division of labor) so that each person can *specialize* in the performance of one job at which he or she is particularly good. The results are synergistic in that we gain more of everything for everyone, a higher quality in all goods, and a lower cost than we would have if everyone tried to do everything for themselves. But specialization is accompanied by certain shortcomings. There is a price to pay for the preponderance of goods and services and progress that specialization allows us to enjoy. That price can be in the form of boredom and often creates an alienation of people from their environment.

A separation comes about with specialization in which people are no longer intimate with the things that make their lives what they are. In the field of technology, it creates a separation of those who understand science and engineering and the accompanying disciplines that create our technological base from those who do not, that is, the general public. We can, through becoming a highly technological society, experience a separation of the society into castes as definitive and absolute as any created in history. We become intellectual haves and have nots.

The computer revolution has dramatically brought home this point. We are already divided into those who are computer literates and those who are computer illiterates. A very large sector of the population does not comprehend how a computer works or how to use one. And the separation is becoming even more pronounced between the younger and older generations where the very basis of logic is shifting from one of simple linearity to one of systemics. This is the same type of shift that occurred with the advent of the printing press when the printed word and the ability to read caused great social upheavals as the common individual's world view shifted from a holistic one to a linear one.

Those who do not understand or cannot use a new technology often feel alienated in their own homes. To the average twenty-year-old a television set is as natural as a tree or a light switch. To a person who grew up in the 1940s, it will never be quite the same device, and the two will never totally understand each other. The same is true of the average individual's understanding of space technology and the scientist's view, with all his or her specialized knowledge of what makes that technology tick. No matter how often a rocket ascends, I cannot help but stand in awe of humanity's accomplishments. Does a scientist or engineer who monitors the spaceship's readouts have the same thrill?

Attendant to this separation is the increase in speed with which we experience change that specialization causes. It is difficult to keep up with changes as they occur. Nor is there time to adjust to the changes. Around six thousand years ago someone invented the plow. The period from the time when that first crude stick was put into the dirt to when the first steel plow was used spans nearly ninety percent of that six thousand years. Such a revolution in agriculture is certainly slow enough for any culture to assimilate the advance in technology. (Though it should be noted that when Peter the Great, Czar of Russia, tried to introduce modern plows to the serfs of his kingdom, they politely accepted them, thanked his representatives, and then burned them, refusing to change their traditional ways.) The McCormick reaper appeared in the latter half of the eighteenth century. The steam tractor soon followed, then the gasoline tractor, and finally the huge combines used on farms today. In each case, the time between major changes in the technology was drastically reduced.

Other technologies show the same pattern. Again and again we see a constant reduction in the time between one major development and the next. Our world is not only changing, but the rate at which it is changing changes. We progress and achieve new levels of sophistication at a higher and higher rate of speed. The result is less and less time to assimilate the changes as they take place and a rising resistance in the individual approaching overload. Do we specialize because of the rapidly increasing mountain of information available to the society? Or does this deluge of information come from the process of specialization? The answer in both cases is yes. And in the face of ever-mounting pressure for the individual in the society to keep up with rapid change, resistance mounts. Keeping up means higher and higher levels of specialization or more and more time spent relearning the nature of our world as that nature changes. Is it any wonder that there is resistance when radically new technologies are introduced?

THE ROOTS OF RESISTANCE

Five causes of homeostatic reactions to technological change have been cited: fear of the unknown, adjustment to anxiety, psychophysiological restructuring, chauvinistic conditioning, and specialization and the telescoping of time. With this impressive list of causes, it may seem surprising that any technological progress takes place at all. Yet at the beginning of this discussion it was suggested that homeostasis has a useful place in the scheme of things, that it is a natural and necessary part of the process. And this is, in fact, the case.

We operate in an ecosystem that includes every living creature on the face of the earth. We are a part of this symbiotic creature called earth just as we are part of the symbiotic creature that makes up our own culture. And all systems are either in a state of adjustment or they are in a state of balance. Adjustment to changing circumstances is in reality nothing more than a system that has lost its state of balance and is seeking to regain its natural equilibrium state. If you stand at the apex of a seesaw, you can balance yourself, albeit precariously. If someone adds a bit of weight to one side or the other, you must *adjust* your distribution of weight to compensate for the change in the dynamics of the system. That is what we do as we experience the changes in our society. Progress and nonprogress are both involved in balancing this growing system called the human race. Without the slower, more cautious considerations of mature members of the society, what would be the result of the youthful exuberance to charge headlong into the future? And without the experimentation and creativity of our technological minds, where would the growth in our society come from? Both progress and caution are necessary, and nature has given us healthy doses of each to ensure not only our growth but also our survival. The time frame is too short to let survival of the fittest take care of mistakes in a natural way. A counterproductive genetic change can die out before it becomes too generalized. A technological mistake, unfettered by the caution of the society, might drag the entire human race into a debacle before anyone realizes what has occurred.

It is no accident that innovations must overcome the objections of the established society in order to survive. It is a test, a method by which humanity determines the usefulness of the new idea. The Bohemian experiment was short-lived. It disappeared in less than a generation. The same is true of the beat generation. Yet the hippie movement remains with us in its influence on our music and our literature, and is reflected in the increased concerns of our society for the well-being of our fellow humans and the environment.

Many technologies have faced the same test. Teflon was invented long before it was put to practical use. Boolean algebra had to wait for the invention of the computer to find its place in the sun. And the automatic chewing gum-operated eyeglass windshield wiper never made it at all. Resistance to the changes presented in these innovations saw to it that only what was useful and beneficial was allowed to survive.

Homeostasis is a natural phenomenon. It is part of what it is to be human just as creativity and technology are part of what it is to be human. Both are necessary for our survival. The importance of homeostasis in understanding the social consequences of a technology is in determining what can be expected, what directions public and private opinion are likely to take, and in anticipating the causes of resistance that are bound to occur. Only in this way can we hope to understand the process of techno-sociological change as it unfolds.

CONCLUSION

Because of the differential manner in which Homo sapiens evolve, that is, by evolving externally to their physical bodies, there is an increased opportunity for the evolutionary changes that occur to be detrimental and yet unextinguishable before an unsuccessful trait becomes generalized over the entire species. Nature uses long time frames in experimenting with species while seeking the creation of maintenance of perfect balance, the less favorable adaptations of a given organism simply dying out while successful adaptations survive and become dominant. With humankind, the telescoping of time becomes a problem in this respect. To help ensure that Homo sapiens do not allow their adaptations to outstrip their ability to control and select for positive adaptive value, humanity also has the survival trait of homeostasis, that is, a natural resistance to change. In this way, the homeostatic tendency can be used as a governor to ensure that adaptations are not generalized too rapidly and possibly allow for the destruction of the species. It functions as a system of checks and balances, necessary for our survival due to the rapidity with which we are able, as a species, to evolve through the development of technology.

THOUGHT AND PROCESS

1. Have you ever resisted a change in the culture? If you have said no, it is not surprising, particularly if you are of a relatively young age. But think again. Answer the questions below. For every yes answer, you have given yourself an example of your own homeostasis.

    (a) Have you ever resisted a new change in mode of dress, such as a completely new style? Did you change your mind as it became more accepted by the general public?

(b) Did you ever experience anxiety over the entry of a new boss or a new worker into your place of business, not being quite sure how to "take them" until you learned more about them?

(c) Have you ever rejected a radical new automobile design as outlandish or ugly only to change your mind later?

(d) What was your initial feeling about computers?

(e) Would you tend to be cautious if tomorrow you heard that an alien had landed on earth and was conferring with officials of the major powers?

(f) Have you ever felt uneasy about going to a party at the home of someone you did not know well?

(g) Have you ever worried about a blind date? Even after you had met the person and could not find anything immediately displeasing about the individual?

(h) Have you ever reacted with dismay at a friend's new hairdo?

(i) Have you ever resisted playing a new game with someone who knows it well?

(j) When you go out to dinner, do you resist trying new or exotic foods, preferring instead the old standbys?

(k) Would you react with caution to a person of the opposite sex entering a job traditionally held by a male or female within your own life? For instance, would you feel uncomfortable working for a female boss or having a male housekeeper or secretary?

(l) Would you react with fear or anxiety if your job were replaced by a machine and you had to be retrained?

2. For each of the questions above, classify the type of homeostatic reaction cited in terms of the five causes mentioned in the chapter (that is, fear of the unknown, adjustment to anxiety, psychophysiological restructuring, chauvinistic conditioning, specialization and telescoping of time).

3. All of the reasons for homeostasis given in this chapter center on fear. How does just good common sense as a reason to resist change fit into this process?

4. Suppose a method of supplying free energy to everyone on an unlimited basis were to be discovered tomorrow. What would be your concerns about the availability of such a technological breakthrough? How would you go about the process of making the technology available to the world, assuming that you had the power to determine the method by which it was to be done (or not done, if that is your choice)?

5. Can you think of any modern examples that parallel the Luddites other than the countercultural examples offered in the text? Do you know anyone with exhibited tendencies toward this form of reaction to change?

The Printing Press
and
Its Consequences

By
Don Steele
and
Robert Burchfield*

INTRODUCTION

There were books long before the printing press was invented, but they were not in the form that we are familiar with. The ancient books or scrolls of the Chinese were handwritten on scrolls or codex and stored by the many scribes during the eighth century A.D. As early as the eleventh century, mold printing existed, and as late as the fourteenth century, cast metal type was known to exist in Korea. By the middle of the fifteenth century, in a period of religious awakening, there was a growing demand for books. To meet this demand, scholars and scribes joined together in the copying of books. Eventually, book factories employing fifty scribes could be found in Italy, Germany, France, and Holland, in which the scribes worked diligently to produce books. It was unfortunate, however, that these books were carelessly composed during the transition. It is often said that need necessitates invention, and it is a period of time in history when there was a growing need to produce vast numbers of books. Keep in mind that it all didn't happen overnight, but the spread of its success overtook the printing technique of that period.

Scientists and scholars disagree on who deserves credit for the invention of movable type, but many seem to give credit to Johann

*Robert Burchfield and Don Steele, "The Printing Press and Its Consequences" (unpublished paper presented at DeVry Institute of Technology, Atlanta, Georgia, August 10, 1984).

Gutenberg as the man who perfected the concept of movable type, earning him the title, "father of printing." Ironically, the ending of one life marked the beginning of a new idea, that of printing. After Gutenberg's death in 1469 and for the next two decades, the new art of printing spread throughout Europe like a raging fire out of control. By the end of the century, more than 40,000 different books had been printed ranging from Greek classics to arithmetic.

Again, as in the days of Gutenberg, demand created a climate for invention. Gutenberg's simple wooden press had survived until the eighteenth century without any major changes. This would change in 1804 when an English earl, Charles Stanhope, built a press of great size made of iron.

The Stanhope press still did not meet the demand because it was hand operated. Then, in 1812, German-born Friedrich Koenig built a power-driven cylinder press with an automatic inking system. However, the sheets of paper were still hand fed. In a joint venture in 1814, Koenig built a two-cylinder press that was automated and could print one thousand sheets per hour. Then, fourteen years later, the English team of Applegate and Cowper built the first four-cylinder press. In 1847, the American inventor Richard Hoe invented the rotary press which mounted the type directly on the cylinder. This would later be modified by the use of duplicate plates. Still there remained a need for a small automatic press which would take up less space but print faster. In 1858, George Gordon made the press to order: his would be called the *platen press* and would become the standard of the print shop. The letter press as we have come to know it is grouped under three basic types: platen, cylinder, and rotary.

The platen press has two flat-surfaced jaws that open and close as it prints. When open, a set of inking rollers apply a film of ink to the form and at the same time paper is fed to the platen. The platen press is not suitable for printing books or magazines, but is ideal for flyers, brochures, and other circulars.

The cylinder press places the form on a flat bed, which moves back and forth under a rotating cylinder. As the cylinder rotates, the paper is fed to it and held by a set of steel clamps. The paper is then rolled over the form while the bed passes under the cylinder. As the bed moves back, the type form is reinked and the freshly printed sheet is fed onto a pile. The cylinder printer is preferred for the printing of books due to its versatility to print vertically or horizontally as a unit.

The final mechanical printing press is the rotary press. Both the impression surface and printing surface are cylindrical. The rotary press prints from duplicate plates, which are mounted onto the printing cylinder. The importance of the rotary press is that it can allow up to five plate cylinders, thus allowing five colors to be printed simultaneously. This is great for newspapers and magazines that favor this printing apparatus.

RELIGION

Before the idea of movable type on a printing press was established and before the Renaissance, the Church that ruled Europe was in its lowest plunge to decadence. The fifteenth century brought to Europe a revived interest in orthodox religion. Minds were stimulated and were reaching out for new knowledge. The renaissance of piety in the Church influenced worthwhile activities among the priests and monks as well.[1]

Because the mind was once again eager to receive new knowledge, the Renaissance also stimulated education. Books were becoming more easily available and a general desire to learn to read blanketed Europe. Only those people in the very lowest classes were the exception. The upper classes were no longer exclusive in the ownership of books. As people began to read, a popular demand for Martin Luther's translation of the Holy Bible increased.[2]

Religion as it existed then was not a personal religion such as is experienced presently. One's religion was inherited, traditional, and ritualistic in nature. The Middle Ages had allowed the Church to become a hierarchy of power, and it demanded, in some respects, many things of the people it governed. No one ever thought in terms of questioning the Church. A mere verbal declaration from the Church was as good as written law. Because the Church went unchecked by its followers, corruption was widespread, and, prior to the Renaissance, literacy was held strictly by the clergy. The scribes hand wrote the only books the world knew, and great sums of money were spent on personal libraries. Common people had no means by which they could purchase books, and if they did fall into a sum of money and decided to purchase books, they would not be able to read them.

Gutenberg's idea of movable type turned the situation completely around and shed some light on a very shadowy organization. As mentioned earlier, the renaissance of piety and orthodoxy created a desire for knowledge. As Martin Luther was seeking to satisfy this new-found desire for knowledge, he thought he was living a life pleasing to God, because he followed the Church. As he studied the Bible manuscripts, he realized that not only had he not been living as his Bible told him, the Church was not doing so either. Martin Luther continued to un-

[1] Douglas C. McMurtrie, *The Story of Printing and Bookmaking, The Book,* 3rd ed. (New York: Oxford University Press, 1965), pp. 125-26.

[2] *Ibid.,* pp. 327-28.

cover alarming new truths and quickly became one of the world's fore-most theologians and translators of the Bible. He developed into one of the greatest leaders of the Protestant Reformation in Germany. His translation of the Bible went to the presses and into the hungry hands of the common people, who had a great demand for the Word.

Religious reformation was a first fruit of the printing press. Mass production of the Holy Bible allowed the average individual to gain access to something that was held very dear. The almost countless num-ber of denominations of Christianity existing today are further exten-sions of Gutenberg's invention. His idea of movable type gave human beings a chance to think for themselves for the first time since the fall of Greek and Roman cultural heritages as the mainstay of reason.

GOVERNMENT

With the introduction of printing, the governments found that they had a potential disaster to deal with. The governmental structure at the time of the invention of movable type was almost completely controlled by the clergy. This period of medieval uprising and cultural impression formed a people's defense for the hierarchical government under which they lived. The opportunity for secular government was beginning to take form.

With the introduction of the press came books, and with books came the freedom of knowledge. Unwilling to face the undeniable cruelties and public unrest that such an invention would bestow on the heads of state, governments instituted plans to curb the people's appe-tite for books. The governments held public book burnings, destroying books that they found to be heretical in content, but in essence con-taining damaging proof of their tyrannical tactics. But when the public outcry turned to violence, the restrictions were lifted and governments turned to a different point of view, determining to join the people in the face of an inability to beat them. It was around the turn of the six-teenth century, after the death of Gutenberg, that government began to play an important role in the printing industry. Secretly funding town printers and openly accepting what was being printed, the govern-ment found an inside tactical method of baby sitting the presses while controlling what was being printed.

In the late 1600s, the British government placed intricate laws on the Colonial printers of the time. The government allowed itself to pick who was to be a printer, how often they printed, and what they printed in Colonial America. This proved to be disastrous, as printing had given rise to a cause for freedom and the people were sure to find that out. However, all was not lost as the government placed in use its own printing press and techniques. With the new press, the government could print public notices, money, proper documentation, and keep important records in nicely printed books.

In the American colonies, the printing press found uses never considered before. The government learned to use the printing press tactfully to achieve cohesion among the people whom they served. Today the government has its own printing facilities in Washington, D.C., known as the U.S. Printing Office. It is encouraging to see an invention overwhelm even the governmental bodies in its practical applications as well as its ability to enhance the means of mass communication and public unity.

INDUSTRY

Industrially, Europe was fortunate. The idea of mass producing print did not catch the contemporary European industry by surprise. As with any invention, an inventor depends on and makes use of knowledge already established in related areas. This is the only way we are able to see growth in technology. Examples of this idea are evident in many inventions. Stagecoaches were reconstructed to serve as the first railroad cars. The first bodies for automobiles were carriages and buggies adapted to support the mechanical power placed in them. Combustion engines were installed in wooden box kites to serve as the world's first airplanes. Obviously, there is an endless list.

Europe's related industrial arts had substantially progressed to a point where they were able to support the idea of mass production print. There were four requirements awaiting this new invention:

1. Paper was plentiful and easy to process.
2. The proper ink was available.
3. Presses for metal and paper were available.
4. Alloys were available to cast, construct, and mold type.

The absence of any of these would have made the development of printing in Europe impossible.[3]

As Gutenberg's application of the printing press progressed, it became a necessary factor in everyday life. One of the first benefits to industry was the obvious creation of jobs. A new type of business was introduced into the world, and it incorporated and stimulated related businesses as well. Printing has evolved and expanded within itself to

[3] *Ibid.*, pp. 126–27

create other more specific businesses and has boosted employment incredibly. It is not surprising to know that today one newspaper company can have over one million subscribers.

The printing press has also made industry more efficient. Take, for example, the basic contract. If every contract, whether financial or otherwise, had to be hand written, where would the consistency and efficiency in management be? There was a method created through the printing press that allowed procedures and policies of a company to be equally delivered to all employees, customers, vendors, and so on.

We have the printing press and Gutenberg's application of it as a major factor propelling us into the future. We no longer live in the Industrial Revolution, nor are we a major industrial nation any longer. We are postindustrial and are currently building an information society to live in. Because of this fact, information must be easily obtainable. We are approaching the point of being able to put our hands on any information whenever, wherever, and however we desire. This process was initiated by that primitive printing press hundreds of years ago. As each year goes by, printed information becomes more easily obtainable, thanks to a process begun with Gutenberg and his press.

SOCIAL

It may seem that in the middle of the fourteenth century, society was in a state of darkness and despair. But to the contrary, it was a time of intellectual activity. Culminating forces in human affairs which had been slowed during the Dark Ages gave rise to the great rebirth of knowledge, the Renaissance. The Great Schism had brought about deterioration of churches, abbeys, and schools and a great falling off in the devotion, enthusiasm, and discipline of the clergy.[4] Minds were once more becoming eager and inquisitive, and an era of exploration dawned as new opportunities were sought for advancement, restoration, and artistry.

Giving rise to a reborn society, the printing press had given the people an opportunity to restructure society, to regain the power to visualize known and familiar facts into new relations, and to apply them to new uses. For example, the textile industry adapted inking techniques to the textile printing industry.[5] Printing shops expanded the industry to include paper mills, metal clerks, pressmen, inkers, en-

[4] Irving B. Simon, *The Story of Printing* (Hudson, N.Y.: Harvey House, Inc., 1965), p. 20.

[5] Douglas C. McMurtrie, *The Book*, p. 125.

gravers, delivery boys, book binders, and many others. It is no great wonder that this early communication tool was so vital to the survival and very existence of fifteenth-century society. The printing press gave rise to the concept of freedom to think, talk, and of course print the truth, no matter what the consequences. We can see the rise of a social-istic bartering system in which one can find businesses growing rich as they cornered the market on goods that had been printed. Not only did the business structure grow, but also the religious society.[6] Regaining trust in the Church by means of the translated Bible which had been printed in several languages was a major step in the restructuring of religion during this period, giving rise to new Church ideologies and de-nominations. The printing press was spreading its cultural shock to the far corners of the earth. One could experience the excitement as hun-dreds of people sought to spread knowledge and carry the art of print-ing throughout the world. Their accomplishment was the satisfaction and livelihood of the printing press today. Society was ready for a change, yet it had no concept of what was ahead in terms of the inven-tion and the legacy of the creation of movable type and the printing press.[7]

The social implications of the printing press are far too numer-ous to document here because they reached every aspect of life at the time of its birth, giving rise to a new social structure, perhaps creating the upper, middle, and lower classes as we know them, or perhaps cul-turally feeding a starving humanity that might have degenerated in superstition, restlessness, and boredom. One might say that society was awakened to a concept of applying existing knowledge in a new fashion, giving way to the great Industrial Revolution as a result of the spread of new or rediscovered ideas through the print media. The print-ing press allowed society to apply itself, to learn about other cultures, and to become specialized in the ideology of learning and experiment-ing. It is from this ideology that a powerful influence, to change and develop, would grow and prosper to the point of actually restructuring the social and economic scale. It is this restructuring effect of the print-ing press that has made this particular invention so vital to the begin-ning of humanity's striving for excellence in the areas of invention, education, and knowledge. Socially, the printing press has left a mark of opportunities for humankind to advance into the realms of the un-reachable goal and to achieve a society that works together, using exist-ing knowledge in an attempt to advance the conditions of a society in which the only road is progress.

[6] *Ibid.*, p. 188.
[7] *Ibid.*, p. 198.

TECHNOLOGY

Our technology has moved us into an information society. The fact that the percentage of gross national product attributable to heavy industry is ever declining while our unemployment percentages are dropping as well is evidence that we have yet to find a way to measure what our nation is really producing. Using industrial output as a measure of prosperity is an archaic approach, developed in the industrial era and no longer appropriately applicable to conditions in this country. The United States produces information and the methods previously used to determine the gross national product cannot be used to determine the amount of information produced and sold in an information society. We can produce information in any way, shape, or form desired. We are also printing it faster every day. It is obvious that the printing press with movable type that Gutenberg designed has evolved into the computer that is so much a part of our lives. Tactical support specialists in the military can load the memory banks of their computers with so much data that page printers will send streaming arcs of printed pages across the room on command, and another single command can destroy the sum total of that information instantly.

Not only is information speedily produced, it is also miniaturized. Microfiche methods have transformed hundreds of volumes of magazines into hundreds of five-inch-by-eight-inch cards. This allows libraries to hold more information without the need to build larger buildings to house the incredible number of books and other printed matter available today. It is conceivable that libraries will become smaller and yet be capable of holding more information due to miniaturization techniques in use today. In the future, a library the size of a bedroom could hold the information of ten of today's libraries.

The ability to print information has been an important and necessary factor of society since the invention of movable type by Gutenberg. It is becoming more important with each passing day. Printing information has allowed one person to build on top of another person's ideas and thus propel the technological revolution forward at an ever-increasing rate. For the information society we live in and are developing, printing is an essential resource. The ability to print is taken for granted for the most part, but without it society as we know it would soon come to a screeching halt. Douglas McMurtrie said it in no uncertain terms:

> In the cultural history of mankind, there is no event even approaching in importance the invention of printing with movable type. It would require an extensive volume to set forth even in outline the far-reaching effects of this invention in every field of human enterprise and experience, or to describe its

results in the liberation of the human spirit from the fetters of ignorance and superstition.[8]

The future of the printing press is speed and miniaturization. It will continue to be the basis for an information society. When engineers design a faster printer, future engineers will read their published documentation and use it only to build still faster printers. The evolution continues. . . .

BIBLIOGRAPHY

Fisher, Leonard Everett, *The Printers*. Lexington, N.Y.: Franklin Watts, Inc., 1965.

McMurtrie, Douglas C., *The Story of Printing and Bookmaking, The Book* (3rd ed.). New York: Oxford University Press, 1976.

Simon, Irving B., *The Story of Printing*. Hudson, N.Y.: Harvey House, Inc., 1966.

[8] *Ibid.*, p. 136.

CHAPTER 3

Creativity and Innovation: The Critical Link

There was once a village, located deep in a hidden valley in Bhutan, in the Himalayas. And this village was located at the very base of a high mountain that loomed over it like a gigantic tiger's tooth, its wall sheer and unscalable, its white-capped summit beyond the reach of any traveler. It was, in fact, known as the Mother of Mountains, the greatest monolith in all the world, and the villagers were proud of its name.

And to this village there came a traveler, a wandering pilgrim of great age and great wisdom. And he stayed a while with the people of the village, visiting with this family or that, exchanging his tales of far-off places and strange sights for food and shelter.

When he had been with the villagers for three days, the pilgrim stayed with a merchant, a man named Mahjidi, who dealt with all of the merchants of the south, and enjoyed the company of strangers more than most. And as they ate, the pilgrim turned to his host and asked, "Mahjidi, I have studied the great wall of stone that you call the Mother of Mountains, and I have seen what appears to be a long staircase cut into its rock face, yet it leads to nowhere. Can you tell me of it?"

"Ah," said Mahjidi, warming to the opportunity to spin tales of his own, "that is a sad and wonderful story indeed. It begins many years ago, when the elders of our village first encountered merchants from the south. They envied these merchants their homelands, rich with the wealth of the world, filled with palaces and great fertile plains, with histories of conquest and stories of adventure.

"And in their wisdom, the village leaders determined one spring day that it would be a valuable thing to be able to reach the peak of the Mother of Mountains and look out over the world from its very roof. The towns of far-off kingdoms must surely be visible from such a height. The sight of all creation spread before them would be a treasure well worth the effort to obtain. The village, they were sure, would become a powerful place and the object of many a pilgrimage, if only they could reach those far-off heights. Then the world would envy the village, rather than the village them.

"And so, to this end, the elders brought the people together to begin the great task of reaching the summit of the great mountain. One person in five was recruited for the task, all their needs to be seen to by the others while these selected few attempted the impossible feat of conquering the Mother of Mountains.

"Through the spring and into the summer the selected villagers strived to reach their goal. They labored long and hard, constructing scaffolding to raise them to the small crevice that ran through the sheer cliff of the mountain's base, some 300 meters above the village's tallest tower. And from there, they began the laborious task of cutting stone steps into the living rock, weaving their way along the crack.

"It was a good plan, they all agreed. It was a plan worthy of their efforts and likely to bring them success. But it was a plan doomed to failure. By late summer, they had barely reached the ledge and begun the steps. By early fall, they had expended the energies of one-fifth of the village on the project, inflicting hardship on all and injury to many.

"'We must press on,' the village leaders would say. 'We must conquer this mountain, and we shall.'

"And press on they did. Through the winter months, they braved the cold to mount their attack on the crevice and slowly extended the long staircase toward their goal. Villagers took turns with the work, no longer relying on only the strongest among them. Everyone was brought into the campaign, each taking their place on the face of the mountain, each enduring the cold, and the danger and the pain of the labor.

"And in the spring, when the warm winds began to blow from the east and the sun rose high into the sky once more, the village found itself more than half way along the crevice, having extended their steps up the wall of the mountain by another 500 meters. Many had fallen along the way. Many more had exhausted their will on the side of the mountain, yet none had given up the quest.

"Then, late in the summer of the second year, the workers on the mountain made a terrible discovery. The crevice had ended. The shelf in the sheer mountain wall had come to an end, and they had barely begun their journey toward the summit. Above the crevice, the mountain mocked them. The glass-smooth surface defied them to continue.

"Even the elders of the community lost heart. The quest was abandoned. The work was ended. And so, to this day, the mountain stands inviolate, the hand-cut staircase to nowhere being the only evidence of the vain hopes of the village. As I have said, it is a sad tale. We never reached the summit, and never knew what glories await the first to do so."

"It is a sadder tale than you believe," said the pilgrim.

"How so?"

"Because the Mother of Mountains is not the roof of the world. And the sight of a thousand kingdoms awaits no one upon reaching its summit. There is only the sky and the beauty of a hundred mountains, aglitter in the white snow that never fades away. To the south, more mountains and hills and finally rolling plains greet the eye, with rivers coursing like silver ribbons toward the sea. It is but a small piece of the creation that is the world. To the north, the land of frozen giants, mountains far greater than this, is all that can be seen. It is beautiful, but far from the wonder your forefathers perceived it to be, and hardly worth the lives and misery of so many people."

The merchant, Mahjidi, eyed his guest suspiciously. "And how would you be knowing this?"

"Because I have been there."

"No man can scale those heights!" snapped Mahjidi.

"And no man need do so. The far side of the Mother of Mountains is a gentle slope that requires many days to cross, but it kills no one, and it eventually leads to the summit. Has no one thought to look to the far side of the mountain?"

"No one," answered the merchant.

The pilgrim nodded. "A mountain has many faces, Mahjidi, as does all else in life. Sometimes it is advisable to look upon those faces in order to understand the mountain."

It would seem that if the village elders mentioned above had taken the time to view their problem differently, they may have saved themselves a great deal of time and misery. But what has a mountain to do with technology? The tale is instructive for our purposes in that it illustrates a key element in the creative process by which our technological world has come about, that is, the need to see things in a new light. Like the mountain, problems in society have many faces. Unfortunately, it is sometimes difficult to see the hidden ones, the ones that may hold the key to the problem's solution, and so, the problem may remain unsolved for some time. If a problem seems to have no solution, it is often not because there is no solution but rather because those investigating the problem cannot see that solution *from their individual perspectives.* Indeed, there is a need to see things in a new light.

To be creative, to be able to develop new machines, new forms of technology, or new ways of doing things, it is necessary to escape from old modes of thinking. It is necessary to see things in a new light, to view them from a slightly different perspective, or to *hold them in a different context* than the traditional one. What really happened to the villagers in the story is that their perspective on the issue was too narrow to see the true solution to achieving the goal. They were so intent on conquering the sheer wall of the mountain that it never occurred to them to look for a solution on some other face of the mountain. To succeed, they needed a change in perspective.

But where does this come from? Are we all equally capable of making this cognitive switch in our approach when the need arises? What of the tried-and-true traditional ways of doing things? Are they without value to us? Should we simply scrap what we have spent so many centuries learning to use for our benefit and charge headlong into viewing things from a different perspective? Or should we cling to the tried-and-true, already tested ideas of the past?

Actually, we do both. An investigation of an unaltered contextual view of reality is an extremely useful thing for humanity, as has been asserted in the first chapter. It is the way we create our technology. And, as we have seen in the second chapter, this does not necessarily mean that we are going to scrap all that has come before in favor of the new. The key word is still *appropriate*. The issue to be dealt with in this section is not whether to innovate, but rather to ascertain what affects this creative element in human activity and how we can use that knowledge to find new answers to new and old questions.

THE ROOTS OF CREATIVITY

Creativity can be defined as the ability to combine a number of factors to achieve a solution to a problem or to make an artifact that is both novel and useful. It is the ability to rearrange known facts into new patterns in order to develop new constructs useful in accomplishing what needs to be done.

Exactly how creativity works and how this rearrangement of facts takes place in the mind has been the subject of research and supposition for thousands of years. The subject of *a priori knowledge* addresses the concept as an alternative to the existence of prior knowledge without experience. Modern researchers seek its roots and a better understanding of the mechanism by which it operates. How man's mind consciously and unconsciously enumerates and evaluates the millions of possible combinations of factors that may contribute to the solution of a problem and the creation of new artifacts is a vast, fertile field for in-depth investigation. The creative process may be slow and methodical, following a linear process, such as is characteristic of the scientific method, or it may be instantaneous, relying on brilliant flashes of insight to bring about an answer. Neither approach is any less creative than the other. Both methods require a new way of thinking.

Creative thinking, the ability to find unique solutions, is something that is learned as much as it is natural. It is *experiential* in nature. The way a person is taught to view life and the experiences one has greatly affect the degree of creativity that person will achieve. Problem-solving abilities are greatly enhanced by experience. The more opportunities a person has to seek solutions, the greater the likelihood that the individual will improve his or her ability to find solutions. A person who has always done the same job the same way without any outside input on alternative methods will be less likely, on the average, to develop unique solutions within the context of that job than someone who has experienced a wide range of different situations and different approaches to doing the job. There is simply a wider range of possibilities available in the second person's mind from which solutions can be formulated.

MOTIVATION

Risk and Reward

Creativity is related to the concept of *risk*, in that risk acts as a deterrent to behaving in a creative manner. The fear of experiencing loss can reduce the desire to be creative. With each attempt at new solutions, one may risk a number of losses, ranging from a loss of personal esteem and social acceptance to a loss of monetary position or even life itself. The first successful manned aircraft was an exceptionally creative thing. Those who achieved it were aware of the risks involved to life and limb and opted to try it anyway. Risk prevented many other casual thinkers from putting their money, and their lives, to the test to prove their ideas. Risk is an important element in the process. It is one of the reasons that the proverb about necessity being the mother of invention is so easily believed. Since there is always risk involved with being creative, it is not until there is a great need (necessity) that people become aware of the problem and are moved to solve it. Often the risk of *not* solving the problem must assume equal proportions to the risk of seeking the solution before the creative process is sparked.

Reward operates similarly as a factor in creativity. It is the reverse side of the risk coin. Reward offers a carrot for those who would seek creative solutions, motivating them through the promise of a better life, more wealth, more efficient use of resources, or a feeling of accomplishment, to name a few. It represents the search for a desirable positive result just as risk represents the avoidance of an undesirable negative result of not being innovative.

Together these two motivations make up what Freud would describe as the "pleasure principle" of motivation, involving a seeking of pleasure and avoidance of pain by human beings. Technology and creativity certainly classify as means of achieving that "idic" goal.

Hierarchy of Needs

A much more useful and succinct survey of motivational factors present in the creative process is offered by Abraham Maslow in his *hierarchy of needs*. Maslow sought to determine what motivates people to behave in the way that they do and, beyond that, to determine what makes a person balanced, successful, and "happy." Through his research, he developed an inventory of needs that must be satisfied in humankind if a person is to achieve a happy, successful, balanced life. This inventory of needs is arranged in a hierarchical structure, ranging from the strongest and most basic motivations on the bottom of the hierarchy, to the weakest yet most advanced goals toward the top of the hierarchy. The Maslowian system has been found useful and appears in several different forms throughout motivational literature as a basic explanation of what determines people's behavior.

Maslow divides motives for actions into five need categories. Starting with the most basic and strongest, they are *physiological, safety, social, self-esteem,* and *self-actualization.* Each category represents a specific type of need in humankind.

Physiological needs are survival needs. They include most bodily needs such as freedom from thirst and hunger, the need for shelter, the need to continue the species (sexual drives), and other bodily functions.

Safety needs are also survival needs. They represent the need not only to be sated, warm, and healthy, but the need to ensure that these conditions will continue to exist in the future. Safety needs involve security and protection from harm, either emotional, mental, or physical.

Social needs, or *belonging needs*, as they are also called, refer to the needs to belong to and be part of a group or society. Human beings are social animals, gregarious in nature, and require the presence of others for their well-being. The need to become part of a group is a survival mechanism no less important to humankind than to members of a lion pride or baboon troop where cooperation and joint effort ensure survival in a hostile environment. In humankind, this need translates itself into the need to offer and receive affection, acceptance, friendship, and a general feeling of belonging to some group. It is from the drive to fulfill this need that the human race divides itself into tribes.

Self-esteem needs involve internal rather than external satisfactions. Self-esteem has little to do with the opinions of others, being based on a personal feeling about one's self. It is considered a higher-order need involving one's self-respect, autonomy, independence from the control of others, and the feeling that one is achieving. Self-esteem needs translate into such external needs as status in the group, recognition for achievements, and attention from others. With self-esteem the person seeks not only to belong to the group but also to stand out in that group, achieving as high a position within the group as can be accomplished. This is also a basic survival need in humankind, acting as an internal driving force to push people toward their limits.

Self-actualization needs represent the highest-order needs that a person has. They represent the need of a person to strive to be all that he or she can be at any given point in time. Self-actualization involves striving to reach one's full potential, to fulfill one's highest aspirations, and to "grow" as an individual to become what one is capable of being. It is the farthest removed of the needs from the instinctual, animal base of the human psyche.

Consider a primitive human being who is thrust into a wilderness area of North America. This primitive person, alone and naked in the world, first seeks to satisfy the most basic of human needs, the physiological. Our primitive—we'll call him Og—immediately goes about the business of satisfying needs to feel sated by first locating the nearest available stream from which he drinks, and then running to kill a wild beast, say, a small mountain lion, which he then proceeds to eat. Once he is no longer hungry and thirsty, he finds himself a nice cave by the stream and settles in, having most of his basic needs satisfied.

Now it is time for him to consider the next level of needs, safety. It suddenly dawns on him that in killing the mountain lion he has put himself in great peril, particularly since mountain lions tend to hunt in pairs. Although he had previously been controlled by his great hunger, that now seems secondary. Once he is full, safety becomes important, and he proceeds to build a fire to fend off any possible attack from his victim's mate, and checks around to be certain that there is a sufficient supply of food (nuts, berries, roots, game coming to the nearby stream to drink) to keep him satisfied in the future. When this is accomplished, not only has he sated his physiological needs, but he feels safe as well.

This is all very well and good, but he now notices a definite lack of companionship. His cave is comfortable, the stream and local flora and fauna are available to satisfy his basic needs, but there are no other people about, which threatens his need to belong. After all, Og is a very belonging being.

To fulfill this need, he takes a hike to the other side of the valley where he has noticed a small group of wandering nut gatherers and invites them to move into the caves next door. They do so, and he now has a group to belong to. So much for social needs.

What about self-esteem? How does Og satisfy his need to stand out in the group? This is no problem for our enterprising primitive at all. Og simply rips a tree from the ground by its roots, walks over to the biggest, meanest guy in the tribe, and bashes him with it six or seven times, proclaiming himself chief and inquiring whether there are any objections that might exist among other tribal members. Obviously, after a display like that, there are none.

Now Og not only belongs to the group, he stands out in the group.

He is chief, leader, and "head honcho." Everyone respects and looks up to him. Everyone listens to his advice and agrees with him. He gets the choicest bits of meat and accepts the responsibility of leading the group. His self-esteem is assured.

But this is still not enough for Og. He has almost everything he could want now, but there's still something missing. He feels "unfulfilled." He feels that he isn't getting the opportunity to fully express who and what he is. As a result, he wanders off to the back of the cave and indulges in his favorite hobby, cave painting. For hours on end, Og mixes pigment and makes brushes and paddles for applying it to the walls, sketching and coloring in fantastic panoramas the hunts and other events from the tribe's life. Now he is balanced, happy, and successful to his full capacity, all his needs being met.

There is still one element of the hierarchical structure of need fulfillment to be taken into account. Maslow tells us that if a lower-level need is threatened, the upper-level needs seem less important and no longer receive attention. It is only after Og was sated that he decided to look out for his safety. It was only after he had become socialized and was a member of the group that he decided to assert himself and become chief. And the desire to express who and what he was did not become a major factor until all of the other needs in the hierarchy were satisfied.

What happens if there is a threat to a lower need? What if the tribe comes to Og and says, "Listen here, Og. We've been thinking that you sit around back here painting walls while we do all the work. We thought we'd pick a new chief and kick you out." Og responds by bashing a few heads, of course. Or, if they've all got clubs and he doesn't, he may just resign in order to preserve his membership in the group.

As soon as a lower need was threatened, the upper needs were put on the back shelf. This is the manner in which most people in society behave. People operate on a principle of achieving needs and protecting themselves from the loss of achieved need fulfillment. Yet the most highly creative people operate at a level of self-actualization. Does this mean that the only creative people are the ones that are totally balanced? Fortunately for humanity, no. Needs are never satisfied totally. What we seek to do instead is *satisfice* our needs, that is, we determine minimum levels of acceptable fulfillment for our needs and work toward the satisfaction of those needs until that minimum acceptable level of achievement is reached. No one is ever completely safe. There is always some risk involved in being alive. If people wait until they *completely* satisfy the need to be safe before working on any higher needs, they would never get any further than that point. No one ever has a total feeling of belonging. There are too many differences among individuals to ensure complete acceptance and understanding from others. If complete acceptance was a criteria for working on *self*-acceptance, no one would ever get as far as considering the possibility.

Satisficing allows us to work on a need until it no longer pulls at us more strongly than another need. We then work on the one that does pull on us most strongly. Without the satisficing compromise, there would never be any great painters, composers, engineers, scientists, poets, military leaders, presidents, doctors, or practitioners of any other profession that demands the participation of the practitioner on the basis of excellence to the limits of one's abilities.

Any of these needs can motivate a person to be creative, yet the drive for excellence for its own sake, the self-esteem and self-actualization needs, seem to produce more creative solutions than do the risk-oriented lower needs.

In all of the need hierarchy, there is either a motivation of protecting what one has (fear response) or striving to obtain what one does not yet have. Higher-order needs are reached based on the desire to achieve and to accomplish, whereas the downward slide into working on lower-order needs is a fear response designed to protect against loss. It is through progress that one can achieve higher-order needs. It is through fear (threat of loss) that one is forced to concentrate on lower-order needs and shelve the desire for higher-order need achievement.

Creativity can be viewed as the result of (a) innate ability, (b) experiential learning, and (c) motivation. This third factor, motivation, can be viewed as being based on either fear (the need to protect against loss) or achievement (the desire to achieve some purpose or gain something not presently extant). Further, these motivations of fear and achievement can be viewed according to Maslow's hierarchy of needs as a matter of an individual seeking to satisfy one or more of the classifications of needs in the hierarchy, either by protecting lower-level needs already satisficed or by seeking to achieve need satisfaction further along in the hierarchy. This is in relation to the creative nature of the individual. Yet to be considered are the social conditions surrounding the individual and the effect these social conditions have on the performance of individuals in reference to their creativity, or the performance of the society as a whole in reference to its creativity. To understand this, a comparison is used.

CULTURAL IMPETUS: A COMPARISON
OF ORIENTAL AND OCCIDENTAL APPROACHES

To illustrate the ways in which cultural and social patterns can affect the degree of technological innovation and the kinds of technological innovation that take place, a comparison of Eastern and Western (Oriiental and Occidental) technology serves well. There are a number of differences in the ways in which Eastern and Western cultures formulate the individual and societal approaches to the subject of change in general and technological change in particular. There are three major factors to be considered: *cultural restriction, linear thinking*, and *philosophical point of view.* All three of these factors are capable of limiting or expanding the possibilities available to a culture.

Cultural Restrictions

Western culture began exerting its influence on China in the seventeenth century. Yet long before that influence began, the Chinese culture experienced a limited but truly advanced technological development. The list of important inventions first developed in China is impressive.

China is credited with the first use of block printing, and the invention of paper, a necessary adjunct to the use of such blocks. Indeed, the first known Chinese dictionary was created circa 1100 B.C. In addition, the Chinese invented silk, the art of weaving, and the first development and use of astronomical instruments, which took place prior to the erection of England's famous Stonehenge, a neolithic arrangement of boulders designed to predict important astronomical events, including the winter and summer solstices and the vernal and autumnal equinoxes by which the primitive Britains could note the passing of the seasons. While European Britains were piling their boulders together, the Chinese were refining their measurements of the heavens.

The Chinese were also responsible for the first use of the horizontal loom, the spinning wheel, the water wheel, the first mechanical clock, gunpowder, porcelain, and the concept of coastal naval defense, rockets, and hand grenades. This is an impressive list. The Chinese culture was an extremely inventive one, possibly the most inventive prior to the Renaissance in Western Europe. The culture was prime for a technological revolution with the possibility of using their inventiveness to great advantage. Yet the technological revolution in China never came about. It would be more than a thousand years before the initial concepts would be rediscovered or borrowed and put to practical use.

The Western European countries, in contrast, were extremely slow to develop an inventive tradition, plodding through centuries of stagnation and extreme restriction in the availability of knowledge before the blooming of Western technology began at the end of the Middle Ages. The Industrial Revolution of Europe occurred only recently from the historic point of view, well after the founding of the United States. What caused the Western culture to succeed in taking advantage of scientific knowledge and China to be unable to make the same transition?

The Chinese culture is an old one. The first known formal government under imperial control was that of Emperor Huang-Ti, who is known to have established the first Chinese kingdom in 2697 B.C. The first dynasty, the Shang or, as it is alternately known, the Yin, was founded as early as 1750 B.C., and there was already a considerable ancient cultural tradition in place at that time. The society was organized and strictly structured under a long succession of emperors whose elaborate and extensive administrative networks held total control over the population. The society was rigid and traditional. Social position was strictly maintained, placing the citizenry in distinct classes across which there was very little mobility. Under such a tradition, any change was seen as a threat to the stability of the society as a whole and the government in particular. A citizen achieved the status of engineer or scientist not by virtue of ability but by virtue of social rank. An individual became a scientist because the father was a scientist. And his son would take on the position after him by virtue of the same reasoning. So tightly controlled was the structure of the society that the population was solidly locked into doing things in the traditional way. Change was considered a thing to avoid. The very religion of the country, based on ancestor worship, was an expression of the people's veneration of the past, past ideals, and maintenance of the status quo. China was considered to be perfect, while the remainder of the world was viewed as barbaric, backward, and uncivilized. To change a perfect social structure would be unthinkable.

In Europe the conditions were diametrically opposed to this stable, tightly controlled Oriental structure. Prior to the rise of economic mercantilism and intellectual freedom, Europe was a collection of petty kingdoms ruled by feudal lords and loosely controlled by kings. Political boundaries changed frequently. The squandering of resources by an expanding population was staggering, and the political control of the Church, the closest functioning body politic approximating stability, was being challenged from within and from without. By the fifteenth century, with the expulsion of the Moors from Europe, the discovery of the New World, the final perfection of the printing press in 1454, and the dominance of European countries in the Mediterranean following the Battle of Lepanto in 1571, Europe was ripe for its own explosion of inventive development. The difference lies in the attitude of the people and the freedom of the people to act. Change among the European cultures had been constant since the beginning of the Renaissance. The people were becoming acclimated to the possibilities of change. When Gutenberg began publishing in 1454, that single event struck the death knell of stagnation in Europe. Knowledge was the key to technological change in Europe, knowledge and its dissemination among the general population. With the relative freedom to improve one's lot offered by the existence of an already entrenched middle class of independent and mobile workers in the trades and the guilds, a burgeoning of scientific inquiry and practical application was inevitable. It was this freedom to act in combination with the availability of information that allowed the Western cultures to move inexorably toward the Industrial Revolution of the eighteenth century.

Linear Thinking

Another major difference between Eastern and Western cultures lies in the manner in which Occidental and Oriental people cognate. The belief among Westerners that Orientals are inscrutable, their ideas unfathomable, and their actions unpredictable and illogical is no more valid than the curiosity and amazement that the Oriental holds for the ideas and actions of the Occidental. Each culture experiences this feeling of paradox when dealing with the logic of the other. The fact that an Oriental sees nothing illogical or paradoxical about a logically Oriental proposition whereas an Occidental may, and vice versa, is testimony to the differences in the approaches of the two cultures.

Western logic is linear in nature. It moves through a sequence of defined statements that leads to conclusions about the world. Statements of logical analysis in Western cultures are bound together by their cause-and-effect relationship. Each statement infers the next. Facts are combined into syllogistic combinations designed to prove that if one event occurs or if one set of circumstances exists, then another event or set of conclusions must "logically" follow. One moves step by step from conclusion to conclusion, each one building on the preceding one in a web of well-founded proofs toward the ultimate conclusion being sought.

For instance, *if* I drop the ball from the tower, *then* it will fall toward the earth. *If* the ball falls toward the earth *and* there is someone standing beneath it, *then* the ball will hit them. *If* these events occur *and* the ball is four inches in diameter and constructed of iron *and* the tower is forty feet tall, *then* the person who is struck will be seriously injured. *Therefore, if* I drop the ball under these conditions, *then* a person will be seriously injured.

This is the type of linear (syllogistic) logic experienced in Western culture. And it is reinforced to a high degree by the written language. Occidental writing and books are constructed to proceed step by step (word by word, sentence by sentence, paragraph by paragraph, chapter by chapter, and so forth) from an initial statement toward a final statement. The movement is physically left to right, front to back (relatively speaking), and top to bottom. The reader must follow a linear sequence from beginning to end that reinforces the ideas of cause and effect, temporal linearity, and beginning-to-end direction. Any idea or concept that does not neatly fit into this type of logical progression tends to be less easily internalized by Occidentals. Indeed, the very use of the word *progression* indicates the Western predisposition toward viewing the world as a *forward movement* of events (cause and effect) through time.

Occidentals are taught to think linearly. They are taught to expect events to follow events and consequences to result from consequences. Logical propositions can be carried forward from one step to another ad nauseam if one so desires. Every action has a reaction, and each reaction becomes an action, creating another reaction.

The Oriental mode of thinking is quite different. In China, particularly during the Imperial Period, the logical process followed a different path—all elements were seen as merely parts of the whole. This holistic viewpoint, creating the image of the universe as a huge, single entity rather than as a collection of discrete parts, predisposes cognition toward an acceptance of an event or condition without regard to its resultant effect on other events or conditions. The Chinese language itself, with its abundance of characters and highly specialized nuances of thought, reflects this. Chinese characters are pictographs, each one representing a complete thought. They are not a series of sound symbols that are additive to form a given word. There is no cause-and-effect relationship among Chinese characters similar to the manner in which the twenty-six letters of the English alphabet, for example, allow us to change the meaning of the characters completely through a rearrangement of their relationship to one another.

The result of this linearity in Western cultures is that whenever there is a new condition, a new discovery, or a new application of known physical laws, one of the questions that follows is, "What's next?" There is always a "What's next?" awaiting the discoverer, and it is a natural consequence of a linear approach to cognition to ask it. For this reason, Western cultures tend to carry advances to their logical conclusion. There is always one more application. There is always one more idea or use inherent in what was previously done with a principle. The sequence progresses until the value of the next step is less than the cost incurred by taking it.

Oriental cultures do not do this. Their holistic approach tends to incorporate a new discovery into the whole *as it is.* Each invention or technological device is seen as a solution to a particular problem and not as the first step in a long series of applications. The Chinese simply do not consider continually asking "What's next?" to be a valid motivator.

Philosophical Point of View

The holistic thinking of Eastern cultures is further exemplified by the philosophical approach developed in the Oriental world. In China, two fundamental principles were applied to the sciences, including astrology, medicine, alchemy, and other Chinese intellectual activities. The first of these, the idea of *yin* and *yang*, presents the universe in terms of a fundamental dichotomy of opposites. Rather than the good-evil dichotomy used in Western culture, in which good must triumph over evil, yin and yang see nature as merely seeking a balance, all chaos coming from an absence of this balance, and a movement toward balance as the way to serenity in nature and in humanity. Balance is the key in the process. Opposites are seen as neither good nor bad, but simply as opposite ends of the scale. Some of the major dichotomies are heaven and earth, male and female, action and passivity, and so forth. Technological change was seen as a means of reachieving balance when there was none. If nature appeared to be in balance, then there was no need to change. This Oriental point of view brings home once more the concept of the preference for the status quo and the veneration of the traditional rather than change.

The second concept that so influenced the Oriental approach to science and technology was the principle of the *five elements*. All reality was classified in terms of five elements, and all spatial relationships and temporal relationships were understood in terms of these five elements. The five elements were *fire, wood, earth, metal,* and *water.* This is somewhat similar to the Greek classification of matter into the elements of earth, fire, air, and water. To the Chinese way of thinking, the efficacy of these five substances as universally elemental in the makeup of physical reality was absolute and inviolable. There was no thought of changing the concept or altering this approach to the understanding of the real world.

All science and all events were described in terms of the five elements and in terms of the balance of the yin and the yang. The Chinese adhered to this theory, unlike Western scientists, who put aside the traditional Greek approach of four elements in favor of alternative theories that more closely fit observation. The Chinese explained irregularities that were observed as examples of the fact that, upon occasion, even the universe goes astray. Astrologers particularly displayed this refusal to accept evidence. As an example, the five elements were thought to be associated with the five planets. In the face of the discovery of Uranus, the Chinese astrologers simply ignored the inconsistency.

In the West we see a striking contrast, specifically during the time of the Renaissance and subsequent political, social, and scientific upheavals. In Europe, with the spread of available literature for the masses and an increasing degree of literacy among the influential, the opportunity for exposure to new ideas arose. Coupled with a cause-and-effect orientation, the population began to question the traditional, to look for new explanations and possibilities, and to turn their attention outward toward the largely unknown and unexamined world around them. The whole idea of academics changed. Papers were no longer written, they were "published," that is, made public, a phenomenon that would have seem ludicrous and without value before the print media had come into general use.

Westerners were willing to look for new solutions, to ask new questions, and to seek information just for the love of knowledge itself. They were relatively unfettered by political restrictions, the information was more available than ever before; and the desire to profit from new technology, coupled with audacity (a "marriage" proven by the success of the rising merchant middle class), combined to push a "next-step" mentality in the West.

The Chinese were eminently practical in their thinking. An invention had a purpose. A discovery was a discovery and it represented a new piece of information that fit into the whole of the fabric of the universe, or it was ignored as a fluke. The culture was inventive, it was organized, and it was highly developed, but the very structure of the logic, philosophy, and social order of the Oriental culture kept it from taking advantage in the long term of knowledge as it developed. In contrast, the immediate reaction to a discovery in Western cultures was, What can it be used for, and what does it mean? What is the next step? What logically follows? It is this linear aspect of Western thinking that gave birth to the Industrial Revolution in Europe once all of the physical and intellectual requirements were met. A similar revolution didn't occur in the East until much later, with the introduction of linearity into Oriental thought. It should be noted that in the twentieth century, this condition has reversed itself. The industrialization of Japan and the strides made by China in recent years are evidence of the revolution in thought and physical conditions wrought by the "what next" linearity in the modern Oriental approach.

THE SCIENTIFIC METHOD

No better single example of linear, logical Western cognition exists than the *scientific method*. It examplifies the step-by-step, proof-oriented approach to the cause-and-effect world that the Western mind embraces. And yet, surprisingly, it begins not in knowledge but in faith.

The scientific method depends on the faith of the investigator at the outset of any scientific investigation for its development. The first action or first step in the process involves developing a hypothesis, an assumption of a real-world phenomenon, and then accepting that hypothesis tentatively, pending proof by experimentation or investigation. The scientist is, in essence, taking the truth of the hypothesis on faith when the investigative process begins.

This is a necessary preamble to scientific inquiry. In the face of limited knowledge, certain assumptions must be made. Based on these assumptions, the scientist then proceeds to test them, thus proving or disproving their validity. Few phenomena are more linear than this scientific approach of observation, hypothesizing, and testing. We *begin* with faith in the Western approach, tentatively accepting observation in conjunction with the proven information available to us. However, rather than stopping there, as the Oriental might do in fitting all observations into a well-ordered belief system, the Westerner *tests* the hypothesis before declaring it to be a true representation of the nature of the world. A key factor in Western science is that there are no absolutes, that "scientific laws" are under scrutiny and are accepted if not disproved, and that we are free to give up a preconception about the makeup of the real world if a better explanation comes along.

Most representations of the scientific method list a set of five steps through which the investigator discovers the truth of a problem. They are

1. **Defining the problem.**   It is first necessary to have a clear idea of exactly what the problem is in order to choose a direction of investigation and properly formulate later experimental techniques.

2. **Observing the evidence.**   An accumulation of information is necessary as an initial step in investigation to ensure that all previous work on the problem is known to the investigator and that he or she clearly understands the nature of the problem under investigation.

3. **Forming the hypothesis.**   The investigator uses intuition and logical thinking to discover perceived patterns of behavior from the preliminary data and draw tentative conclusions from those patterns. This is a primary creative step by which a new way of viewing the nature of the reality under investigation takes place.

4. **Experimenting.**   A method of testing is designed to either validate or deny the truth of the hypothesis. The design of the experiment itself is a crucially creative process. The nature of the experiment will automatically exclude and include what can be discovered by its results.

5. Formalizing the theory. The results of the testing of the original hypothesis and conclusions are made available to the public (published) for the scrutiny of the investigative community as a whole and to use to further investigate the subject.

Again, note the step-by-step linearity of the process. The scientific method lends itself to creativity and innovation both in the sciences and in the implementation of known concepts in practical technological application. To form a hypothesis about a problem requires creativity. To design an experiment to prove or disprove a hypothesis requires an innovative approach. To view science and the scientific method as not creative because of their dogged requirement of proof before acceptance is to miss the point. True creation of useful technology requires this rigid approach, but within an orientation that admits to the possibility of change and that encourages the usefulness of innovative methodology to improve efficiency and the quality of life in the society. Change for its own sake, as we will discuss later, is an equal confusion of issues, but change for the betterment of the population is valid, natural, and desirable.

On the negative side, the scientific method by its nature may pose as great a threat to innovation in the society as that of restrictive social norms. The five-element approach of the Chinese is no more than a given point of view regarding the nature of the structure of the universe. That is equally true of the accepted "modern" viewpoint. Modern scientific theory represents a different viewpoint, but it is still only a viewpoint. Westerners use the scientific method to discover new facts within their world view. Each hypothesis is built on the basis of observations and beliefs that support that accepted world view. Because of this, their ideas tend to support the traditional body of knowledge and what it expresses. It is only with great difficulty that the major paradigms of a culture are changed. It is short-sighted to believe that the use of the scientific method alone can bring about a discovery of unique and final answers to questions of science. Too many external influences serve to taint the investigator's objectivity. Social, political, and economic motivations sway opinion and prejudice choice of experiment. It is safer to agree with accepted, dominant scientific beliefs than it is to break radically new ground. The kinds of conclusions we reach as a result of our observations tend to follow tradition. People have an incredible capacity to pigeonhole experiences into neat packages that fit in with their present view of the world. Radical departures are often ignored or rejected as "illogical" or "unrealistic." Anyone who has seriously considered the informational content of the general theory of relativity will tell you that it defies common sense. That is one of the reasons that its acceptance was so very long in coming, even in the face of experimental proof.

A radically different scientific theory can only be accepted if there is a change in the world view of the scientific establishment. People must be willing to change their ideas about a subject and be open to the possibility of an alternative interpretation's validity. This does not come easily.

Herein lies the main problem. Experimentation tends to follow accepted theory. Experiments are committed to theorizing, instrumentalizing, conceptualizing, and developing methods of problem solving that are bounded by the present paradigm. Without a willingness to diverge from the generally accepted paradigm, the resolution of problems whose solutions lie outside society's present realm of knowledge is difficult if not impossible. Many a discovery has been made in ignorance of accepted scientific beliefs, which, if the accepted beliefs had been known, would not have been possible. Yet it is only by changing its world view that a society can progress. If the answers to major scientific problems lay within the present belief systems, they would have already been solved. The scientific method can be a valuable tool, and just as with any other tool, it can also be misused. The question is not one of supporting or destroying the present paradigm. It is a matter of discovering the truth of our physical environment and how that truth can be used to society's betterment. As long as this is the key element of research, the scientific method itself will maintain its usefulness.

CONCLUSION

Technologizing is a creative process, and the nature of creativity and innovation by which technological progress takes place is affected by the culture and social structure within which that change is occurring. The manner of cognition, the sociopolitico-economic structure and the philosophical makeup of the population can all either detract from or encourage creativity and technological development. The direction that scientific inquiry takes and the manner in which a culture puts scientific knowledge to practical use are dependent upon the *belief systems of the culture*, the *opportunity* and the *ability to technologize*, the *motivation* to do so, and, most importantly, the *freedom to question* the established concepts of the culture. Without these elements, technological progress does not take place.

THOUGHT AND PROCESS

1. The following suggestions were presented by students during a brainstorming session, designed to release their creative bent in determining the solution to an industrial problem. The problem given them involved making a useful product

from a relatively useless by-product, in this case, coffee grounds. Scan the list and choose an example that you consider to be logical and one that you consider illogical. For each of these choices, develop a rough plan of how to produce and market the product you have chosen. It may surprise you how logical some illogical ideas can be when thought of in a creative manner.

Products from used coffee grounds: (a) fertilizer, (b) records, (c) chewing gum, (d) ashtrays, (e) concrete filler, (f) make weaker coffee, (g) products with a coffee aroma, (h) dye, (i) pigment for paint, (j) pet food, (k) sopping up oil spills, (l) soap, (m) an abrasive, (n) a protein substitute, (o) hog food, (p) fermented for fuel, (q) make disposable garbage cans, (r) make particle board, (s) fire extinguishers, (t) make a song, (u) ballast, (v) insulation.

2. Test your own creativity. For best results, it is suggested that you try this question before reading the next one. For those of you who have just resisted that temptation, this should be a most effective exercise. Place a single piece of paper in front of you on a desk or table and place a pencil beside it. With no other materials to work with, *be creative!*

3. If you have had a problem with the preceding question, it may be from programming. Many people have the belief that they are not creative and therefore are incapable of doing something creative. To bypass this, try doing question 2 again with the substituted instruction—make something. After all, anyone can make *something,* even if it is only a mess!

4. Innovation is nothing more than finding a better way to do something. Choose an everyday job that you dislike but must perform and be innovative in discovering a new way to accomplish the same task. Creating an artifact is not a stipulation of this exercise, yet if you can incorporate such a physical construct into the process, more understanding will be gained.

5. Go sit under a tree. How do you view the world around you? Do you analyze it piece by piece or view it as a whole? Try to follow your thought processes as you look around and take in your environment. Whether you find yourself to be linear or holistic in your approach to viewing the world around you, take a few minutes and try it from the opposite point of view. Now try this with other areas of your life and see how it increases your ability to be creative, innovative, and able to solve problems.

CHAPTER 4

Economics and Cultural Impetus

One of the most potent determinants of technology among the subsystems of our sociopolitico-economic system is in the area of economics and economic structure. Technology is an integral part of economics, being one of the chief causes of success and failure, not only for individual enterprises, but for the economic system as a whole. What is it about technology that affords it such a major role in the performance of the economy? Is it really such an essential part of the economy's functioning? What would be the effects of less or more technology from the economic point of view? These are some of the questions that are explored in this chapter.

CAPITALISM: THE TRADITIONAL APPROACH

In order to understand the place of technology in the economy, it is necessary to first explore the nature of the economy as a whole, with particular emphasis placed on technological progress and its resultant effects in the economy.

Economics can be defined as the study of how societies choose to use scarce productive resources for the production of goods and services and how they choose to distribute those goods and services to the general population for their consumption. Inherent in this definition are the primary elements of any economic structure, specifically, *production*, *distribution*, and *consumption*. In brief, economics studies how goods are produced and distributed, and indicates that the purpose of all this activity is the consumption of the produced goods. (*Consumption* is the utilization of goods and services to produce satisfaction in the user. It is not within the scope of this book to delve into an understanding of exactly what is meant by producing satisfaction. Since the concept of satisfaction is such a personal, subjective experience, that is one particular "bucket of worms" that the author gladly passes by.)

The decisions that are made regarding how the production and distribution are to take place are as numerous as the economies that exist today. We shall concentrate on the modern economic approach used in the United States, which is a "mixed" economy.

The roots of the modern American system of production and distribution lie in capitalism, a theory of economic structure first introduced by Adam Smith in his book, *An Inquiry into the Nature and Causes of the Wealth of Nations*. Published in 1776, Smith presented a basic explanation of the economic structure of "modern" nations and how that structure functioned.

Characteristics

To explain the how and wherefore of economies, Smith started with certain assumptions about how people behave in the market and how the market structure performs its major functions. These initial premises were true in 1776 for the environment within which Smith operated and with certain modifications are largely true today. Some of his assumptions were as follows:

1. Private property.   Adam Smith stated that one of the characteristics of a successful and naturally functioning economy was adherence to the principle of private property. According to Smith, individuals hold ownership of the means of production in the form of *natural resources* (virgin land with its original fertility and mineral deposits), *labor* (the potential and real contributions of people to the production of goods and services through work, which involves both physical and mental effort), and *capital* (all manufactured productive resources such as plant and equipment, machinery, and improvements made to land that render it more suitable for the production of goods and services). He further stated that the property owners have the right to do what they will with that property, whether it be to use it (consumption), sell it, rent it, or do nothing (save it). As such, the owners of property and other productive resources are free to determine to what use that property will be put, and what goods and services, if any, they will be used to produce.

2. The principle of self-interest.   Smith expressed his assumption that all individuals operating within the economy carry on purposeful activity designed to further their own self-interest. Smith saw economic action as a means of bettering one's position in life and as only having that purpose. The idea of self-interest, Smith contended, may seem somewhat cynical and negative on first inspection, but upon deeper reflection, it can be seen, not as something to be criticized, but as something that is a fact of life. Technically, Smith noted, there are no actions, even altruistic ones, that are not done for the purpose of self-interest.

This "idic" Freudian approach to economics is nothing more than a statement of the pleasure principle which holds that individuals tend to seek pleasure and avoid pain. Far from being in opposition to the religious and philosophical beliefs of Smith's era (he was living in a Scottish environment during the age of Puritanism in Great Britain when the Protestant ethic was dominant), the concept is quite a valid one. As presented, Smith merely reminded the reader that the desire to improve one's economic welfare (to be better off) is the reason people interact in the market and involve themselves in the economic processes. Why would someone want to work if they received no reward for their efforts? What would be the motivation to create new products or increase the efficiency of one's job performance if the result was not of some benefit to the person? Even those who enter relatively less profitable professions receive benefits, either to the best of their personal ability or as a combination of monetary reward and personal satisfaction. The motivation for doing anything is to improve one's quality of life. This desire for self-interest in the economic sense translates into the concept of the *profit motive*, that is, the concept that the reason for people involving themselves in economic activities is that they believe that they can receive a profit (when revenues exceed costs) by doing so, and therefore improve their economic position. In fact, it is assumed that the economically rational person always seeks to maximize his or her profit.

3. Competition.   The definition of economics offered at the beginning of this section dealt with the *choices* available to individuals regarding production and distribution of goods and services. The necessity of choice concerning what will be produced and what will not be produced reflects the fact that resources are *scarce.* There is not enough land, labor, and capital available to satisfy *all* objective desires, and we must therefore choose which economic desires will be taken care of and which will not be taken care of. As individual firms and individual consumers, all with different ideas about what is and is not an important use of resources, we must *compete* for the right to use what resources there are. The functioning of the market system is such that this competition works to our advantage by forcing the society to produce only that which is most desirable and only in the most efficient way; to distribute it to the largest number of people; and, in the process, to maximize the amount of profit we are able to accumulate. Paradoxically, even those who fail in this "survival of the fittest" sort of competitive mentality end up economically better off as a result of the process. However, to see this more clearly, it is necessary to deal with the concept of competition in conjunction with another principal characteristic of capitalism—consumer sovereignty.

4. Consumer sovereignty. Who makes the decisions in a capitalistic system? Who decides what will be produced? Who decides who will receive the manufactured goods and in what quantity? Who decides what prices will be and in what quantities goods and services will exist? According to Adam Smith, it is the consumer who makes all of these choices, and it is because of this that the market system is self-sufficient, self-regulating, and self-correcting. The scenario that results in this conclusion develops in this way.

In an economy made up of many buyers and many sellers, each of whom holds a relatively small part of control over what is bought and sold, producers are at the mercy of the market. They have no control over the prices or over their own individual success, and they cannot individually affect the activities of the market. They are operating in what is known as a perfectly competitive market. Suppose a company is producing pencils and is in competition with many thousands of other pencil manufacturers. Such a manufacturer must compete with other pencil makers for the right to scarce resources such as pencil lead, rubber for erasers, and wood for the casement into which the pencil lead is fitted. The pencil makers also compete for labor, for machinery and equipment, and for the money necessary to set up and run a business. In the product market, our pencil manufacturer is faced with the problem of selling the company's product to the public, something that all the other pencil manufacturers are also trying to do. In the face of such competition, how can the pencil manufacturer get ahead? How can the company's economic welfare be improved to the greatest degree by maximizing profits? Since profits are the difference between total revenues, that is, whatever is brought in from the sale of pencils, and total costs, or whatever is expended on resources to produce the pencils, the manufacturer must either have a higher price (to make revenues higher) or a lower cost (to make total costs lower) or both, so that the difference between total costs and total revenues is as great as it can be. (Total profits equal total revenues minus total costs.)

Case 1: Raise Price.   By raising the price, the pencil company should reap higher total revenues, thus raising profits if costs remain the same. However, since consumers have so many choices as to where they can purchase pencils, they would be foolish to buy at the higher price when so many other companies are selling the identical product at a lower price. Our pencil manufacturing company loses *all* of its business when it raises the price and will make *no* profits rather than more profits. Thus it cannot raise the price and survive.

Case 2: Lower Price.   This seems like a viable solution. If the company loses all of its business to other companies when it raises its price, perhaps it could take all of its competitors' business by lowering its price and thus increasing total profits by selling more items. In this case, however, our pencil manufacturer is faced with a new problem. As soon as the company lowers its price, its competition will see the threat to its business and will drop its own price to meet our company's price. The result is that everyone is selling about the same number of pencils as they did before, but because the costs are the same and the price is lower than before, profits are less. Hence the pencil manufacturing company will not improve conditions by lowering the price any more than it will by raising it. It is stuck with the price that the market offers, that is, the price that *consumers* are willing to pay.

Conclusion: Consumers set the prices in capitalistic markets, not manufacturers.

Case 3: Increase Production.   What about just making and selling more pencils? If the manufacturing company can do that, it can make more profits without changing the price. Can it do this? Not easily. In order to sell more than the company's equal share of the pencils that are demanded, the manufacturer must have a product that people want to buy *more* than they want to buy someone else's product. That is, the pencils must be better in that they must present more utility to the public. The manufacturer will strive to make a pencil that is as good as possible in order to encourage the public to buy it. Unfortunately, so will all of the competitors, with the result that *all* of the pencils that are available in the marketplace will be of the highest possible quality. If one company has a better product, the other companies lose customers, and they are not about to let that happen. If they do, they go out of business and are no longer part of the market.

Conclusion: Consumers determine how much of a product is going to be demanded in the marketplace, and they further force manufacturers to produce the highest quality product that can be made, given present levels of technology and present levels of expected customer utility.

Case 4: **Reduce Costs.**   Another possibility is open to the pencil manufacturing company if it wants to maximize profits. It can lower costs and thus increase the differential between total costs and total revenues. This is done by using every technological and human relations skill at its disposal to minimize production costs—by searching for better production methods, by buying resources at the lowest possible cost, and by generally keeping expenses to a minimum—to maximize the efficiency of the operation and to waste nothing. This tactic works to maximize profits, but, unfortunately, it works equally well for the other manufacturers with whom it is in competition. All the pencil manufacturers strive to be as efficient as possible in methods of operation. The result is that the manufacturer must maximize efficiency just to stay in business.

Conclusion: In order to maximize profits, competition forces a company to produce the highest quality goods it can, at the lowest possible cost, to sell them at the lowest possible price that allows a profit, and to make them available to the largest number of people.

Consumer sovereignty exists not only in the product market where final goods and services are bought and sold, but is equally evident in the factor markets, where natural resources, labor, and capital are purchased.

Earlier it was mentioned that these same manufacturers who were trying to maximize profits in a product market were also in competition for productive resources. In order to get a share of those resources, they must be willing to pay a high price for the right to their use. The factor market is an auction from the point of view of the manufacturer. It consists of bidding against all other producers for the right to resources. Only the manufacturer who is willing to pay the highest price will be able to share in those resources.

Where does that money come from? It comes from profits generated in the product market. If a company is unsuccessful in selling its products, it will not have funds to buy raw materials for the production of more goods and services. Only successful sellers are capable of being successful buyers in the factor market. And how does the economy decide what products are bought and what products are not bought? *Consumers* do it.

Each time a product is bought, the consumer has voted for the right of that product to exist. Each time a consumer refrains from buying a product, he or she has chosen to vote "no" to the right of that product to exist. Only the products with the most votes (those that are sold the most, creating the most revenues for their producers) are allowed to exist. "No" votes result in no money, which results in no product.

Conclusion: Because of competition and consumer sovereignty the market creates the largest number of goods of the highest quality for the lowest price and makes them available to the most people, with the final consumer having the final say in what will be produced, how it will be produced, who will produce it, and to whom it will be distributed. The market system is therefore democratic, self-perpetuating, self-regulating, and self-correcting in nature.

5. Capital investment.   Adam Smith stressed the importance of investing in the means of production. One of his clearest messages in the *Wealth of Nations* was the belief that *economic power stems from the ability to produce.*[1] It is only because of a nation's capacity to produce that it has economic strength, and the maintenace of that ability ensures a viable economy. He further felt that investment should be encouraged and that it is through an expansion of the capital base, that is, the level of investment in the means of production (plant, equipment, and so forth) that an economy can experience economic growth and an improvement in prosperity. As will be illustrated later, it is the validity of this conviction that brings us to the actual value of technology in the economic structure.

Laws of Supply and Demand

The importance of these two primary laws of market operations lies in the fact that they are central to both predicting the actions of suppliers and consumers in interaction and in explaining why prices in the market reach the levels they do and why they tend to maintain those levels in the short run. Simply, the laws may be stated as follows: the *law of demand* states that there is an inverse relation between the price of a good or service and the amount of that good or service that will be demanded in the marketplace. The *law of supply* states that, in general, there is a direct relationship between the price of a good or service and the amount of that good or service that will be supplied by producers for sale in the marketplace.

These laws of economic behavior are an expression of common sense, based on the observation of consumer and producer behavior in market situations. They are, to a considerable degree, self-evident. As buyers and sellers, virtually everyone who has interacted in the marketplace has a personal experience of their validity.

Law of demand.   The law of demand states that if the price of a good rises, fewer units of that good will be demanded by the public because fewer people are willing to pay that higher price, because those

[1] *An Inquiry into the Nature and Causes of the Wealth of Nations*, Adam Smith, 1776.

who do purchase the product purchase less of it, or because of both of these factors. Hence, people ask, "How much?" before deciding to buy. Price is a determining factor.

A simple illustration of the law of demand is seen in graphing the relationship between changes in price and changes in demand. (See Figure 4-1.) Note the downward slope of the demand curve indicating the inverse relationship between the price of the product and demand for the product in question.

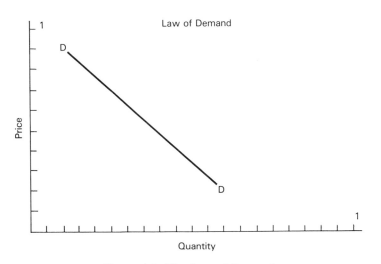

Figure 4-1 The Law of Demand

Law of supply.   The law of supply is similar in that it expresses another facet of market behavior: the determining nature of price in deciding how much producers are willing to produce. As we indicated above, individuals seek to maximize profits in their production of goods and services. The greater the profit that can be gained by producing a good or service, the greater the probability that a producer will be willing to produce and offer that item in the marketplace. Hence, if the price of a good rises, producers will make more of it, as it represents an increase in the contribution each unit sold will make to their economic welfare.

Figure 4-2 illustrates the direct relationship between price and quantity demanded. The supply curve slopes upward to the right, indicating an accompanying rise in quantity supplied for each increase in unit price.

In the case of the law of supply, it should be noted that there are some exceptions. The wording of the law of supply begins "In general . . ." This qualifying phrase is included because of three cases in

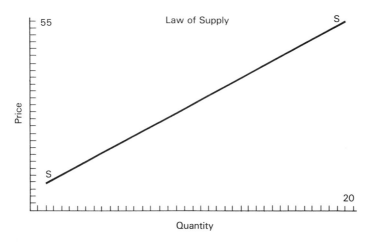

Figure 4-2 The Law of Supply

which the law does not hold: the case of *fixed supply*, of *fixed price*, and of *economies of scale*.

The case of fixed supply is illustrated in Figure 4-3. Here we are faced with a market in which, no matter what the demand for the product or what price consumers are willing to pay for the right to possess that product, the quantity of the product available remains the same. There is no relationship between price and quantity supplied in such a case. Examples of this phenomenon include such things as the Mona Lisa, the Hope Diamond, or the quantity of moon rocks available on earth. With each of these products, no matter how much the price changes, the supply is fixed and constant. (It should be here noted,

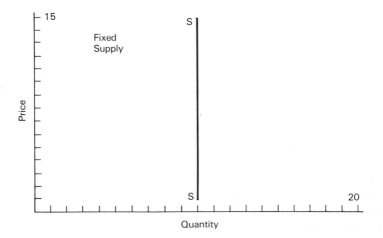

Figure 4-3 A Case of Fixed Supply

however, that this can be a short-term phenomenon. In the case of the moon rocks, if the price were to go high enough, someone would mount another expedition to recover more. The price must exceed the cost of the journey for this to happen, however, which is not likely in the foreseeable future with the present levels of technology.)

The second exception, the case of fixed price, is also indicative of a market in which there is no relationship between price and quantity supplied. In this case, however, it is price that remains constant, not fluctuating at all over a wide range of quantities of product supplied. The supply line becomes a horizontal line (see Figure 4-4), since quantity is free to vary as much as it wishes while price stubbornly refuses to budge. Price controls imposed by a government or cartel are an example of this. If the price of gasoline or the price of airline tickets is fixed, computed by a predetermined method independent of demand for the service, it is a case of fixed price and results in a breakdown of the law of supply.

Figure 4-4   A Case of Fixed Price

The third exception, that of economies of scale, is a bit more complicated. The supply curve in this case slopes downward to the right rather than upward as we would expect (see Figure 4-5). It *seems* to indicate a situation in which as the price of the good drops, producers are willing to make more of the item. Such an idea is contradictory to everything we know about producer behavior. To illustrate, it would seem to indicate that if the going price of washing cars declined from $7.50 per car to $2.50 per car, more people would be interested in doing it! How can this be?

The fallacy lies in the nature of what is taking place in the market. Rather than describing changes in supply in response to a change in

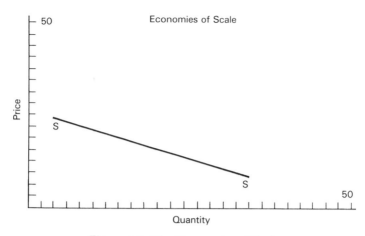

Figure 4-5 The Economies of Scale

price, the graph describes the relationship between changes in *cost* and changes in price.

As the number of units produced increases, manufacturers find that their overall costs tend to decline. This is due to an increase in the efficiency of plant and equipment as production rises. As an example, consider the information provided in Table 4-1.

If our pencil manufacturing company purchases a pencil-making machine at a cost of $100,000, that amount of money is added to its costs of production. If it makes one pencil with the machine, that pencil will cost it the price of the raw materials, the cost of one worker to push one button one time, and *$100,000 in machinery cost per pencil!* In order to make any profit at all, the pencil would have to be sold for a price in excess of $100,000! However, if the machine has a capacity of producing 1,000,000 pencils before it must be replaced, and if used to do so, the pencil company has now greatly reduced its machinery cost from $100,000 for one pencil (a cost of $100,000 per pencil) to

TABLE 4-1 DILUTION OF FIXED COST AS PRODUCTION RISES

Fixed Costs for Machinery	Units Produced	Fixed Costs per Unit	Left for Profit @ $5.00
$100,000	1	$100,000	($99,995)
100,000	5	20,000	(19,975)
100,000	10	10,000	(9,950)
100,000	100	1,000	(995)
100,000	1,000	100	(95)
100,000	50,000	2	3
100,000	100,000	1	4
100,000	1,000,000	0.10	4.90

$100,000 for 1,000,000 pencils, or a cost of $0.10 per pencil. In order to achieve this improved level of efficiency, all that is required is an increase in output. This means that as a manufacturer increases production, the *cost per unit* declines, resulting in higher profits.

This is all very well and good, but why would our pencil manufacturer want to lower the price if this was true? Why not just keep the price the same and get more profits per pencil sold as costs drop per pencil produced? *Because in order to sell all of those extra pencils, the manufacturer will be forced to reduce the price.* The law of demand states that to increase demand for a product it is necessary to reduce the price. So what does the company do? It lowers the price as it increases production and still makes more profit as long as the costs per unit decrease faster than the price per unit decreases. Hence, the peculiar downward-sloping nature of the supply curve.

Market Equilibrium

The laws of supply and demand, as an expression of consumer behavior and producer behavior, offer an explanation of how quantities of goods made available in the marketplace and demanded by consumers differ with a change in price. The curves represent the collection of price–quantity combinations one would expect to encounter in the marketplace. However, in and of themselves, they do not tell us what *will* be produced and what *will* be purchased by consumers in the marketplace nor what the actual price will be. It is necessary to view the behavior of both consumer and producer in concert to discover the actual price and the quantity bought and sold. This is called *equilibrium.*

Equilibrium exists in the marketplace when buyers and sellers agree on both the price of goods bought and sold and on the quantity of goods produced and sold at that price. If agreement is not reached, there is no *prevailing price* in the market, making demand and supply uncertain. The *equilibrium price* is defined as the price at which suppliers and consumers agree to exchange an identical quantity of goods or services and agree to go on exchanging that quantity at the same price. As in Figure 4-6, if the supply and demand curves are constructed on the same chart, it can be seen that the same combination of price and quantity exists on both the supply curve and the demand curve at only one point, that is, where the two curves cross. It is the only set of (x, y) coordinates (quantity and price) shared by both consumer behavior and producer behavior patterns. Graphically, this is the point of equilibrium, where the supply and demand curves intersect. The price prevailing at this point is known as the equilibrium price, and the quantity demanded at this price, the equilibrium quantity, is the quantity that will exactly match consumer demands with producer desires. Any other price will create instability in the market.

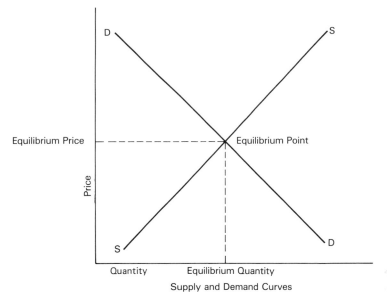

Figure 4-6 Supply and Demand Curves

Suppose that for some unknown reason the price was actually higher in the marketplace than the equilibrium price. If this were the case, there would be a difference between the quantity demanded (at a higher price, demand would drop) and the quantity supplied in the marketplace (at a higher price, the law of supply dictates that there would be more produced). The resulting difference would mean there is more product available than there is demand for, and suppliers would have leftover supplies after they had sold all that they could in the market. This excess product costs them money. They have purchased resources to produce these goods, they have their capital tied up in them, and they must pay for the warehousing of the produced items while they are still in the suppliers' possession. The suppliers are very interested in getting rid of the surplus, and the only way to do that is to sell it, which means convincing the public to buy more than they have of the item by lowering price.

As this process takes place, the dropping price makes goods more attractive to the buyer, who is now willing to buy more; whereas suppliers, no longer able to command such a high price for their goods in the marketplace, are less interested in producing the goods. Thus supply drops toward the equilibrium quantity level. Eventually, equilibrium price is again reached and quantity supplied once more equals quantity demanded, resulting in a minimum of waste and a maximization of efficiency.

Result. If the price in the marketplace is higher than the equilibrium price, there will be surplus product, which forces producers to lower the price, thus disposing of merchandise as demand rises, forcing price back toward equilibrium.

What if the price is below equilibrium price? At a price lower than equilibrium level, manufacturers will not be so ready to produce, since they cannot make as much money for their efforts as they can at a higher price, and they will cut back production. Hence, supply drops. At the same time, the lower price will appear very attractive to consumers, who will demand more of the product than at the higher price, and demand will rise. This results in *shortage* in the market, a situation in which demand for the product is higher than the supply.

Under such circumstances, consumers find themselves all seeking a relatively limited supply of goods, and they must *compete* with one another for the right to own those goods. Such competition on the part of consumers is called an *auction*. The consumers bid up the price until one person wants the item enough to pay a higher price than anyone else is willing to pay. Bidders drop out as the price rises; at the same time, the higher price results in suppliers supplying more of the product for consumption. Demand drops and supply rises as price rises, until, at equilibrium, demand and supply are once more equal.

Result. If the price is below equilibrium price, a *shortage* of goods results, forcing a rise in prices through the auction process until equilibrium price is again reached, supply is exactly equal to consumer demand, and there is no waste of productive efforts.

One of the major benefits of a capitalistic free market, according to Adam Smith, is exactly this tendency of the market to always seek equilibrium. It means that the market system is self-correcting, that it is not necessary to control it, to second-guess it, or to force it to perform in an efficient manner; this is something that happens naturally.

Economic Trade-Off and the Production Possibility Curve

In the *Wealth of Nations*, Smith points out that resources are often scarce. There are not enough of all productive resources to fulfill all subjective desires for goods and services. Although this may not be strictly true of all resources (the problem may be realistically viewed as one of distribution rather than true scarcity, in that a person residing on the shores of Lake Superior may argue against the idea of there being a shortage of water, whereas a citizen of Egypt or Libya would not hesitate to agree), the economy behaves as if there really is scarcity and behavior is, after all, what we work with in economics. It can be assumed, then, that we have effective scarcity, and as a result, we are forced to make choices among the various uses to which we can put our resources. Every hour of labor put into producing automobiles is an hour not put into the production of water buckets. Every ton of steel used in the production of railroad cars is a ton of steel not dedicated to the production of skyscrapers, or subways, or M-1 tanks, or battleships, or any of the other choices available. Not only are we forced to make such choices, either overtly or through the market system, but we are forced to pay a price for the decisions we make. This concept of the *opportunity cost* is the foundation of the phenomenon of economic trade-off and the resulting production possibility curves.

An *opportunity cost* is the value of the alternative *not* chosen. It is the value of what was given up in order to do something else instead. If a person has the opportunity to either go into the restaurant business at an expected profit of $20,000 the first year, or into the tire business at an expected profit of $32,000 for the first year, but is unable to do both, then a decision must be made. The economically rational human being is expected to seek maximization of profits and would therefore choose to enter the tire business and earn a profit of $32,000. An accountant would consider the profit of the tire business to be exactly $32,000, the amount earned after expenses the first year. But not so to the economist. To the economist, this approach would ignore the cost of giving up the opportunity to make $20,000 by entering the restaurant business. Thus the economist would consider the profit made *because of choosing to go into the tire business* as only $12,000, since the individual would have made $20,000 no matter which business he or she chose to enter. He or she simply absorbed a loss of $20,000 resulting from foregoing the opportunity to enter the restaurant business, and the *economic profit* was $12,000, the difference between the total accounting profit and the opportunity cost.

Translating this concept into a general case concerning economic structures, each time goods are used for one purpose, it costs us in that we are forced to pay an opportunity cost for not using the resources for the second-most profitable purpose. We *trade* the right to use it for one thing in favor of the right to use it for another. This *economic trade-off* (a sacrifice that must be made in order to obtain something) is inherent in each and every economic decision.

Economic trade-off can easily be demonstrated if we consider the case of an economy that is forced to choose levels of two goods that compete with each other for some or all of their inputs.

Imagine a society whose production capacity is limited to making either bicycles or canned soup. It can either put all of its productive capacity into making bicycles, or it can put all of its productive capacity into making canned soup, or it can make various combinations of bicycles and canned soup. Each of these cases involves a trade-off.

The production of nothing but bicycles results in no canned soup being produced and vice versa. If we decide to produce *some* canned soup, we must give up *some* bicycles in order to do so, as every case of canned soup requires resources that would have been used to make bicycles, thus reducing the number of bicycles that can be produced. By plotting all of the various combinations of bicycle production and canned soup production that are possible, we create what is called the *production possibility curve*. (See Figure 4-7.)

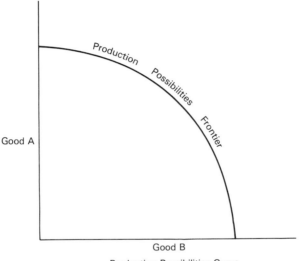

Production Possibilities Curve

Figure 4-7 Production Possibility Curve

Note that any combination of production that lies on the production possibility curve itself (known as the *production possibility frontier*) represents a case of full utilization of productive capacity. It defines the *maximum* combinations of output of the two products, given that all resources are fully employed. It points out that if we want to produce an amount of canned soup, x, then the maximum number of bicycles that can be produced at the same time will be an amount, y, that corresponds to the point x, y on the curve, given the present level of resources and a constant technology. Any combination of bicycles and canned soup lying beneath the production possibility frontier is possible but represents less than maximum use of resources. Any combination outside the production possibility frontier is a combination that is impossible as it is beyond the capacity of the society to produce. The *production possibility frontier* is the limiting expression of possible production combinations of the two goods due to economic trade-off in the economy.

Although there are no societies in the world today whose productive expertise is limited to the production of either bicycles or canned soup, the concept of economic trade-off and the production possibility curve is still quite a valid one and, as we see shortly, one that is intimately involved with the problems and benefits of technology in the society.

Capital Formation

Smith stresses the importance of capital formation throughout his exposition of capitalism. Indeed, the name *capitalism* itself stems from his assertion that investment in the means of production is essential for a successful economy. He states that, unlike the assertions of the mercantilists who viewed economic power as resulting from amassed physical wealth, true economic wealth stems from the ability of the society to produce goods and services. It is this ability that brings economic prosperity, growth, wealth, an increase in economic welfare, and all of the other positives of a capitalistic system. Investment, he says, is the key. Without an investment in the ability of the society to produce, machinery and equipment would soon wear out and there would be no replacements; production would thus stop. The society would stagnate and slowly begin to drift backward toward less productive times. Efficiency in the use of resources and the availability of productive ability keeps the economy going. Investment is essential.

To a large degree, all of these characteristics and principles are as true today as they were in 1776, yet we find some major differences in classical capitalistic theory and modern capitalistic approaches. One obvious example is the tremendous contributions to the economic system made by government at all levels, especially at the national level. Smith stated that the only legitimate role of government in economic affairs was to protect property rights and that its role should be limited to just that. We do not seem to have followed this precept too closely.

Why? What is different now? Why do we find the traditional capitalistic theories so inadequate to explain modern economic phenomena? Primarily because of an occurrence that took place after Smith's theory was formulated, an occurrence that totally changed the complexion of economic structures. What changed everything so dramatically was the Industrial Revolution.

CAPITALISM AND TECHNOLOGY: THE MODERN APPROACH

In the industrialized societies of the twentieth century, the classical approach to economics presented by Adam Smith is no longer adequate to explain what takes place in the market. We no longer live in the simplistic agrarian society that surrounded Smith in eighteenth-century Scotland; all that was put to rest by the advent of the Industrial Revolution and the rise of industrial states. The level of industrialization necessary to create and maintain an industrialized nation is far beyond anything that Adam Smith had envisioned. Progress is more rapid; the range and availability of goods is many times greater; and the productivity of the worker due to improvements in technology, methodology, and communications has exceeded that of Smith's time so completely that his approach is simply inadequate.

This is not to say that the Smithian approach to economics was incorrect or that it is now useless. Because of the industrialization of a considerable portion of the planet, circumstances have changed so much that his theory is no longer reflective of actual conditions.

The effect of the Industrial Revolution and its consequent explosion of technological change is similar to the effect of quantum theory on traditional physics. Sir Isaac Newton successfully predicted the results of a wide range of combined events through his classical physical laws. His laws of gravitation, action and reaction, and so forth were quite sufficient to explain observed phenomena in his time. However, as the phenomena studied by physicists changed, and specifically as the phenomena increasingly involved particles traveling close to the speed of light, the results of the observations diverged from what the classical laws said should be happening. It was necessary to wait until the beginning of the present century for Einstein, Bohr, and others to formulate new laws before the observations could be fully understood. The Newtonian approach to physics was not wrong, it was simply inadequate to explain phenomena occurring at relativistic speeds.

Theory was not able to closely approximate observation again until the advent of major breakthroughs in economic theory, such as the Keynesian approach, developed by John Maynard Keynes. In this century alone, economic thought has shifted from a patchwork of classical theory to the Keynesian approach, then to the emergence of the Monetarist school, and finally to what is known as the "modern economic approach." The adherents to the latter approach do not seek to decide whether Keynesians, Monetarists, or other theorists are correct in their assumptions, choosing instead to concentrate on a more pragmatic approach and use whichever theory works at a given point in time.

Technology and Efficiency

In the classical approach, it was indicated that individual manufacturers compete for the dollars of the consumer by producing as good a product as they possibly can. This is still the case. However, there are now other forms of market systems in operation in our society beside the traditional highly competitive free market. There are oligopolies, markets made up of large manufacturers of goods, each of whom is able to command a significant percentage of the market. Industrialization allows such a market to exist.

Why would a market exist with a limited number of producers? What happened to the large number of small producers, each with a small part of the market? They have given way in many areas of endeavor because of the insatiable desire on the part of the economic system as a whole to operate more efficiently. *Efficiency* is the ability to operate with a minimum of waste, effort, and expenditure of resources. In industry, that often translates into bigness.

Oligopolies have come about because of the fact that as industrialization takes place, efficiency is created through large-scale operations. Economies of scale dictate that in order to produce a good product at a low price, it is necessary to make many of them, at least until the process becomes too large and cumbersome and the benefits received from bigness begin to decline. Since, as previously noted, industrialization seeks bigness in order to achieve efficiency, if a market is limited by the size of the consumer sector of the economy, that is, the number of consumers and the number of units each wishes to consume, the resulting production profile shows that only a few large companies are needed to satisfy all demand for a product. The market still operates competitively, with each company doing its utmost to achieve maximum production at minimum cost to create the best product it can and sell it to the largest number of people at the lowest price. Yet this is now achieved through a small number of relatively large companies in whom is concentrated the lion's share of industrial capacity and not in a large number of small companies operating without benefits of economies of scale. We call this phenomenon *industrial concentration*, and the greater the degree of concentration, the more highly oligopolized that particular industry is said to be.

Thus we have a small number of automobile companies, a small number of steel companies, a limited number of major computer manufacturers, and so forth.

This does not rule out large markets. In some cases, bigness can lead to complacency. When that happens, inefficiency is created. The economic system, being a self-correcting system seeking balance, has an answer for this. When the imbalance becomes great enough, new producers are encouraged to enter the market, albeit an expensive process, and the inefficient producers are forced to return to a competitive status or lose business. This was classically illustrated by the influx of foreign automobiles into the United States from foreign producers who were able to produce at a lower cost and compete favorably for consumer dollars, in spite of protective tariffs and stiff resistance from domestic producers. The American auto industry finds itself in a position of having to increase efficiency in order to compete.

How does a company increase efficiency other than through economies of scale? Again, by the use of technology. Technology is what allows bigness in the first place, but that is not enough. Industries are constantly seeking new methods and better ways of handling production problems, and they are constantly seeking to develop new products for the consumer. Technology is a means of doing both. Money spent on the research and development of products and on improving production methodology is as much a part of the attempt to compete as it was in Smith's time. Only the highest quality and most useful products at the lowest possible prices are able to survive in the market for any length of time. And that means improvements in technology.

Economic systems, particularly capitalistic economic systems, are reciprocal in nature. They offer a return on the basis of value. That is, people are recompensed for the value they contribute to the economy. How much are people paid for their services? It is generally dependent on how useful their work or service happens to be. The more valuable the input into the productive process, the more recompense they are able to command in the marketplace. Their value is, of course, in proportion to the availability of workers doing the same type of work. Scarcity alone does not produce value. Buggy whip lacers, for example, are extremely rare in our society, but that is not a guarantee of a high price for their services, as the service being offered is not particularly valuable to the society in itself. In contrast, an electrical engineer in a world of electrical gizmos, gadgets, and goodies is a valuable resource, and if there is a limited supply of these skilled professionals, the value of each one is relatively high. What is valuable commands value in return for the right to use it.

In addition to scarcity, productivity creates a high price for the services of workers. The more productive one is, the more valuable one becomes. Likewise, productivity is created through technology. In an industrialized society, which is dependent on the efficient use of knowledge, machinery, equipment, and so forth, it is through technology that we maximize the value received from the production process. The only reason for technology, as indicated in the first chapter, is to create artifacts based on natural law for the purpose of doing something more efficiently and with less effort and expense. Labor becomes more productive through the addition of technology to the productive process, and the wielding of that technology in its efforts to create goods and services.

If an economy is to produce goods and services efficiently, it must have the technology to do so. Technology becomes a limiting factor on the capacity of a society to create economic welfare for its members. This brings us to technology and the production possibility frontier.

Technology and the Production Possibility Frontier

Societies make many choices and the economic trade-off of each contributes to the nature of the society. As an example, the political form associated with a particular economic construct is dependent in part on the choice that a nation makes between creating and distributing goods through the private sector (industry) and creating and distributing them through the public sector (government).

Any good may be either a public good (provided through the governmental structure) or a private good (provided by private industry and made available to the public through the market system). There is nothing inherently "public" or "private" about any given good or service. This is strictly a matter of choice. The deciding factor is whether the society feels that a particular good can be more efficiently (more economically) provided through one sector of the economy or another. Private armies, for instance, are not unheard of, but they are not nearly as efficient and economical on a large scale as a public defense system organized and paid for by the society through the public sector. Housing is generally created and supplied by industry in the private market, but when it is necessary to build houses for the good of the society and the private industrial sector is unable or unwilling to do so, the government (public sector) is not averse to stepping in and creating that housing through the process of eminent domain and law. This is simply a *choice* that is made by the society.

Another choice that the society makes is whether to invest productive capacity in the ability to produce (building capital inputs such as machinery and equipment, new knowledge through research, and so forth) or to create consumer goods (shoes, ships, sealing wax, and electronic computers). This choice determines, among other things, the rate at which an economy grows and the economic welfare of a society over time.

Even as Smith said, it is necessary to invest in the means of production. A society must set aside part of its productive capacity for use in producing capital if it wishes to have the means necessary to supply that society with goods and services. The choice is how much to set aside.

Investment in technology is part of that decision. Technology dictates the total productive capacity of the society along with other factors, such as the availability of resources. How technology is handled affects economic growth. A society can choose to invest any portion of its wealth in capital, and the more it invests in capital, the greater the future capacity to produce all combinations of goods and services. It is through this investment that the production possibility curve is able to be extended. If productive capacity is increased to more than compensate for depreciation and depletion, then the result is *more* capacity to produce than before, Therefore the production possibility frontier moves outward. Since the area beyond the production possibility frontier is defined as an area of impossibility, by extending the curve, we decrease what we cannot do, and increase what we can. Economic welfare rises with an increase in the availability of goods and services, and economic growth is experienced.

Determinants of Economic Growth

Economic growth takes place in accordance with a mix of determining factors. As these factors change, so does the ability of an economy to grow. The major accepted determinants of growth include:

1. *Population:* The size and nature of the population, including demographic factors such as age, sex, health, location, and total numbers.
2. *State of the arts:* The type of technology available (state of the arts), the availability of that technology, and the applicability of it to the production and distribution of goods and services.
3. *Growth of knowledge:* The degree to which knowledge increases in the society among individuals and the highest level of knowledge obtained by the society as a whole, how generally knowledge is dispersed through education, how specific it is to content (the nature of knowledge), and how easily exchanged and communicated that knowledge is, are important factors in what constitutes the intellectual base of a society and the rate at which new knowledge can be assimilated and put to use.
4. *Available resources:* The type and quantity of resources available; the rate at which resources are used; the rate at which various resources, if any, are replaced; whether replaceable resources are an important factor in the economy of the society; and the expectation for future increases in resource availability.
5. *Rate of capitalization:* The rate at which the society is willing to capitalize or invest in productive capacity, within the limits of the other factors mentioned here. The capitalization rate is as much a measure of attitude toward economic progress and growth by the members of the economic community as it is a measure of the rate of change in productive capacity itself.

Three of these five determinants of economic growth—state of the arts, growth of knowledge, and the rate of capitalization—are directly related to technology itself. The state of the arts is certainly a matter of technological development in the society. The rate of capitalization determines how much of society's resources are going to be devoted to improving and using technology for the benefit of the society. Technology as artifacts *is* the investment of the society in capital. The application of what is known to create efficiency in the productive process and to produce an increased availability of goods is central to economic welfare.

The third factor, the growth of knowledge, is generally viewed as the most significant source of long-term economic growth. It is through increasing our understanding of the physical world and then educating the population in that knowledge that we are best able to deal with and manipulate the physical world to society's benefit. It is this that makes education such a critical factor in the future of the country, and it is this that creates the opportunity for a wide range of people to experience the benefits of improvements in technology. Technology results from manipulating nature, an ability that stems from our understanding of its behavior and one that results in an increase in knowledge.

Unlike resources, the quality and quantity of technology and knowledge can be increased in a society. As resources decline in abundance and quality, innovative use of the remaining resources increases the importance of technology as a means of maintaining or improving the state of the economy. Efficiency in the face of diminishing natural resources and a growing population that demands more and more goods, can only take place through ever-increasing amounts of capital, knowledge, and technological application of that knowledge.

Population, too, has a link to technology. The ability to gain new knowledge and manipulate that knowledge to the benefit of society through the creative process to form technological artifacts is a function of the availability of intelligence in the society itself. Assuming that it requires a certain level of intelligence to manipulate new information effectively, the size of the population can be critical.

Genius, as an example, is measured in percentages, reflecting the top 2 or 3 percent of the population in the ability to solve problems. Assuming that this is a valid measure, the number of geniuses available in a given society is proportional to the size of the population. A population of 1 million people would be expected to yield an average of 20,000 people who qualify as geniuses (2 percent). If the population reaches 10 million, that number jumps to 200,000 citizens with genius capabilities. The larger the population, it can be argued, the larger the pool of genius intelligence on which the society can draw. It should also be noted, however, that the larger the population, the greater the number and extent of societal problems to be handled.

As for natural resources, there are only two possibilities available to us for solving the problems of limited, rapidly depleting supplies of raw materials. In the face of these decreasing resources, we must either increase the efficiency with which we use what we have, thus increasing the amount and quality of goods and services yielded from a given amount of input, or we must search out new supplies to support our present level of consumption and increased future desire for consumption. In either case, technology is necessary to accomplish the goal.

Efficiency, as already noted, requires improved methodology. In the case of finding new supplies of raw materials, consider the parallels between the discovery of the New World and the exploration of near space. When the New World was discovered in the fifteenth century, the European countries increased their supplies of raw materials dramatically. In a time when the denuding of the huge European forests had all but destroyed available supplies of wood, the New World was found to have abundant supplies readily at hand. A single huge forest stretched from Canada to Florida and from the East Coast to the far side of the Mississippi River and beyond. Gold and silver, gems, and iron and coal deposits were all available in huge quantities.

Mercantilism, which dictated that economic power stemmed from physical wealth, flourished with the availability of a virgin world to take that wealth from, during what has come to be known as the "rape of the New World."

Later, however, the capitalistic possibilities were realized fully and colonies developed to take advantage of those increases in raw materials. It was no accident that most of the British Navy during the heyday of British colonization was built from American naval stores. Eventually, the process led to the emergence of the world's greatest industrial powers.

Similarly, we exist in an era of decreasing resources. Yet just beyond our reach, in a sea of velvet void, there float incalculable supplies of nickel, iron, and other minerals just waiting to be scooped up and processed by some enterprising group of people. The only factor that has kept this process from occurring to date is the expense involved in comparison with the profits that would be incurred from the enterprise. When costs drop low enough, or demand for those resources rises high enough, a profit potential will result and, just as any economically rational human being would be expected to do, entrepreneurs will venture forth in search of greater economic welfare for themselves and their people. The potential exists for such an enterprise in the near future, and the results could be the birth of economies undreamt of prior to this time.

Technology and Economic Welfare

Economic welfare refers to the quality of life that exists in a society. This is not necessarily synonymous with economic output, nor is it dependent on economic growth per se.

Suppose you were living in a small, agricultural nation with a low level of technology and a large but relatively poor population. Suppose also that the basis of dietary staple in your country is a hypothetical tuber called "blivet." Nearly every meal is centered around some form of blivet, which is chiefly ground and cooked to form a brown porridge with a flavor similar to cardboard. Now suppose that improvements in agriculture are introduced into the country that result in a massive net increase in agricultural output, due to trade agreements, some fortuitous investment from outside the country, and an extremely enlightened government. The increase in output represents growth in the economy with a rise in the gross national product (GNP) from a level of 100 million cabas (the caba is your monetary unit) to a level of 200 million cabas. What a wonderful turn of events! In just one year, your economy has doubled! The country should rejoice in its good fortune!

Or should it? Prior to the rise, blivet represented 60 percent of the GNP of your country. Every family in the country could afford three bowls of blivet porridge per day, a very adequate diet. Now, with a doubling of production, each family finds that it can have *six* bowls of blivet per day! Is this progress? Are six bowls per day of sticky brown porridge with the consistency of cardboard an improvement over three? This is a highly subjective question.

It appears that the economic welfare of a country may be equally dependent on the *quality* of products consumed as it is on the *quantity*. It is the quality of life we are establishing, not just the quantity of goods.

In this regard, technology contributes not only to the efficiency with which a nation is able to produce but also to the variety of goods that are available. Consumers have more to choose from in their efforts to satisfy consumption desires if the level of technology allows for a wider selection of consumption items from which to choose. And through this process, technology contributes heavily to the economic welfare (quality of life) of the members of the society.

Technology and Negative Externalities

It is inequitable to view technology in society only from the perspective of economic benefits. A balanced treatment of the subject must also note ways in which the economic structure suffers as a result of technology. The side effects of technological industrialization are not always pleasant. Industry's use of technology, if done without regard for the total fabric of the world system within which it exists, can lead to the polluting of rivers and streams; the saturation of the atmosphere with deadly substances; the destruction of delicately balanced ecosystems; and the decrease in living conditions through crowding, tension, pressure, and a host of physical and mental disorders in the people that technology is designed to help. Collectively, these negative side effects are referred to as *negative externalities*, and even with the homeostatic tendencies of the species, it may not always be possible to resist economic progress enough to catch these deleterious effects. In truth, it may be a price we think we are willing to pay.

One negative externality is the cost that must be absorbed by some third party outside the exchange process of the consumers and producers, as a result of the exchange process taking place in the market. Yet it should be understood that, in the case of technology, it is the *application* of technology and not the concept of technology itself that causes the problem. It is the way in which we *handle* technology that creates difficulties.

For example, a farm products company locates on the banks of a stream in a rural setting, close to several small towns with a readily available supply of labor and in close proximity to the farming communities to whom it wishes to sell. Being a good businessperson, the owner of the factory endeavors to produce a high-quality product at a reasonable cost and, as part of that effort, chooses to dump low-level toxic by-products into the local stream. The alternative waste disposal process would be so expensive that the price to customers would dramatically increase. In good faith, the company is seeking to keep costs down and to produce a useful product, which it then sells to wholesalers and farmers in the surrounding area. The farmers benefit from a farm products plant being so close at hand, the wholesalers benefit from the easy availability of goods to sell, and the manufacturer benefits from lower production and distribution costs. The consumer of the vegetables and other produce benefits from better-quality food at a reasonable cost. Everyone seems to win.

Or do they? What about the waste flowing into the local stream? If the stream can handle the increase in toxic materials, everything is fine. However, if it cannot, there are negative externalities. Suddenly the manufacturer is no longer seen as a firm being maximally efficient by minimizing costs, but rather as a firm that is passing costs on to a third party. Downstream, the farmers may begin to experience problems with their livestock that drink from the stream, having to deal with low milk production and even illness and death of productive animals. Towns further downstream may experience increases in purification expenses in their efforts to draw drinking water from the water source, purify it, and distribute it to community residences. Plant life may suffer from the chemicals in irrigation water, resulting in erosion, loss of productivity in fields (requiring, by the way, a greater amount of fertilizers and other farm products to maintain yields), or even serious flooding in the area due to heavy losses of plant life. All of these negative consequences have to be paid for by someone. The cost of production did not go away, it was merely *transferred* to the community as a whole.

Does the farm products company really get away with anything? Are its customers really free to purchase the farm products at an artificially low price? Economic systems are reciprocal in matters of externalities as in other matters. The consumer and the manufacturer pay for the disposal of the toxic wastes, but in deceptively different ways.

Taxes in the community pay for water purification. If the farm products company is located in the community, it shares in that tax burden. Its customers also share in it, which reduces the amount of money they have available to pay for farm products, thus reducing the company's sales. Farmers suffer losses in crops and livestock, further reducing their ability to buy from the manufacturer. Loss of prestige and esteem may plague the firm if the negative externality comes to light. The general quality of life in the community may decline. Law suits may result. The employees and principals of the company further suffer from higher taxes and a lower quality of life. In many small ways, the cost of the dumping of waste is paid for, and paid for by the firm creating the problem. As with the other factors in the economy, technology creates a trade-off through negative externalities, exacting an opportunity cost from the society.

CONCLUSION

Technology affects the economic structure of a culture through a number of relationships that exist among social content, cultural characteristics, and the nature of the technology itself. Through its capacity to increase efficiency and raise the level of output of a culture, technology creates the opportunity for the society to grow and develop. Innovation is a key element in the creation (production) and distribution of goods and services to the population so that they can be consumed and create satisfaction. The very concept of economic welfare, that is, how well off the population is in terms of the quality of the life that they lead is linked with the capacity of the economic structure to supply these goods and services and, then, in ever-increasing levels of sophistication and variety. Technological innovation depends on the scientific capabilities of the society itself; the quality and the extensive nature of the society's educational system; and the percentage of the total productive wealth that is invested in the development of research and new methodology, new processes, and new products and services.

On the negative side, for every technological innovation and for every advance in the economic utilization to which a society puts its natural resources, labor force, and capital base, there is an inherent cost that must be paid. Because of the natural law of reciprocity, a culture has to pay that price no matter what. If the effects are not handled logically and with forethought by the society, the payment may be higher than necessary and may actually reduce the level of economic welfare through destruction of the environment, overuse of limited resources, or other unforeseen results of technological activity. For this reason, technology and its role in economic structure should be taken into account in its development and utilization *before* it results in negative externalities beyond an acceptable level and beyond the capacity of the society to absorb.

THOUGHT AND PROCESS

1. What are some of the ways in which technology has changed the work experience for your family in the past ten years? Consider both positive and negative consequences of technological change as they have affected you or other working members of your family.
2. Make a list of technological changes that are presently under scrutiny for their detrimental effects on the environment, on the safety of the members of society, or on the economic well-being of one or more groups of workers in our society. Include some less obvious examples affecting only small parts of the population at present.
3. Assume that through innovative technological change the costs of building houses were to be reduced by 50 percent in the next six months. Speculate about some of the probable economic changes that would take place as a result of the new technology.
4. What changes in the economy are likely to be the result of a general use of industrial robots? What changes have already come about?
5. If a single input into the economic process (natural resources, labor, capital, technology, entrepreneurship) had to be limited to its present level, which would have the most devastating effect on the development of society? Why?

CHAPTER 5

An Idea Whose Time Has Come

From the previous chapter, we learned that technology is a primary element in shaping the economic structure of the world in which we live. Without technological development, we would not be able to increase output in the face of dwindling supplies of productive resources, nor would we be able to increase efficiency or find new methodology for performing economic functions. The wide selection of goods and services available to us today to create economic welfare through their consumption would not be available, and our future would be one of stagnation, dropping production, ignorance, and descent into an increasingly primitive social structure.

This being the case, it must also be true that society as a whole embraces technology wherever it can find it, rushing to utilize the growing wonders of a technological age, striving ever forward to greater heights of production and consumption. But this is not the case. The reluctance of society to embrace new technology goes beyond the simple homeostatic reactions of the population. There is a very real phenomenon in action here, and it is one that follows very natural lines.

To see what that process is, let us first consider the work of an early capitalistic economist and scholar, the Right Reverend Thomas R. Malthus. Malthus was a contemporary of Adam Smith, living and working at about the same time, and his particular interest in economics dealt with the interaction of economic structure and population growth. He is further considered to be the first political economist of the capitalistic age, and is credited with being instrumental in having economics labeled the "dismal science." As we shall soon see, this was not without good reason.

THOMAS MALTHUS AND THE MALTHUSIAN PROPOSITION

Malthus presented a paper in 1798 entitled *An Essay on the Principles of Population*. The essay stirred up quite a controversy among economic scholars. Malthus presented his findings on the rise in population compared with the rise in food supplies, concluding that the human race was doomed to misery and extinction through starvation. His argument was as follows.

Consider what takes place when a rich nutrient medium such as agar is introduced into an agar plate for the study of bacterial growth. An extensive though limited (finite) supply of food is made available to airborne spores that can grow once they have landed on the plate. As this happens, it is possible to study the number and size of the bacterial colonies as they develop and experimentally determine in a short time the effect of growing populations in the face of finite resources.

By observation, it can be seen that there is a difference in the rate of growth of the colonies through the life of their development on the agar plate. In the beginning, a small number of colonies appear as small patches in the agar. They quickly enlarge, however, the size and number of patches growing at an increasing rate until a number of flourishing colonies are present. What follows next is a period of relatively constant growth, during which the number of colonies tends to stabilize and the growth rate remains about the same until the agar plate approaches saturation. The third stage is one in which the rate of growth declines and finally comes to a halt, eventually resulting in the decline of colonies and the dying out of bacterial groups in the plate.

The curve inscribed by the growth rate of the colonies in the agar plate is called a *sinusoidal curve*, because it has the same shape as the sine curve of trigonometry. It is an S-shaped curve indicating first, the growth at an increasing rate; second, the growth at a steady rate; third, the growth at a decreasing rate; and, last, absolute negative growth, or decline (see Figure 5-1).

Malthus reasoned that this is analogous to the conditions facing humankind on a planet with limited resources. Whereas our food production is growing, unlike the steady state of the agar plate nutrient, it grows at an arithmetic rate, that is, it rises in a straight line and can be represented mathematically by a linear equation. The population, on the other hand, grows geometrically, doubling and redoubling in ever-shortening time spans, creating a graphic line that curves upward to the right. By comparing the two growth rates, Malthus discovered

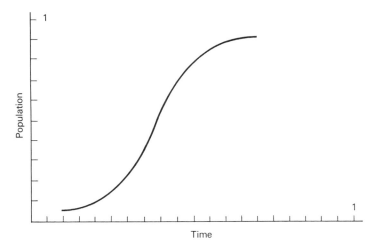

Figure 5-1 Biological Growth Curve

that the growth in population may soon outstrip the growth in food supplies, and when that takes place, the fate of humankind is the same as it is for the bacteria in the petri dish. In other words, according to Malthus, the human race experiences growth in a sinusoidal curve format through time.

What has happened to the bacteria is that they were introduced into an environment with an extensive food supply that is benign and offers the opportunity for growth. Being opportunistic in nature, the bacteria grow rapidly, sending out spores to form other colonies, their growth rate thus increasing at an increasing rate. As the petri dish approaches its maximum capacity to support bacteria, the growth rate slows. Spores are not so easily established on the surface of the dish as before, since it is now full of colonies. In addition, as with any biological entity, the bacteria begin to emit waste products which are poisonous to the organism, and this further reduces the amount of available space and agar for new colonies. Growth continues through this period, but at a much stabler rate. Finally, in the third state, the plate becomes crowded and heavily laden with waste products, producing a growth rate that declines and declines at a faster and faster rate, until the organism runs out of food and can no longer support itself in the face of the waste poisons in the petri dish, and the bacterial growth ceases, being replaced by a lethal decline in the bacterial colonies present.

According to Malthus, as the human population begins to grow faster than the food supply and to surpass it in growth rate, the necessary result will be famine, pestilence, disease, poisoning of the population, and ultimate extinction (see Figure 5-2). Is it any wonder that economics was labeled the dismal science?

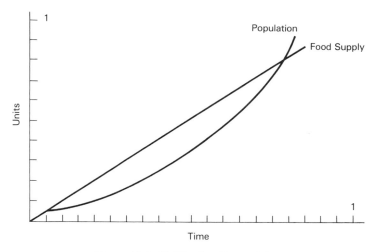

Figure 5-2 Malthusian Proposition

And Malthus was right. There was nothing inherently incorrect in his analysis of the situation. Yet we still thrive and at population levels far beyond those predicted by Thomas Malthus. Where is the difference?

Malthus was correct in his analysis, as far as it went, but he forgot to consider one important difference between humankind and bacteria or, for that matter, between humankind and any other animal. We evolve and adjust to the environment mainly through external means, and at a very rapid rate. We *innovate*, and Malthus failed to take this into consideration.

Early in the history of the human race, it became troublesome to hunt and gather and thrive in the process. Using hunting and gathering as economic formats requires a great deal of land per person. As the population rose, this became increasingly difficult. Rather than enter a flat period in the growth curve, as Malthus's analysis would have predicted, human beings chose instead to try slash-and-burn farming, clearing land by cutting and burning off the foliage and then planting crops to be harvested. Two things happened as a result of this. One was that humans ceased wandering, at least wherever the new technique was used. The other was that it was so successful that there was actually a surplus of crops, leading to trade, leisure time, and ultimately growth in the population. Population rose until food supplies could not keep up with the rate of population growth. Early farmers then switched from slash and burn, a system that quickly wears out the land, to one of plowing and fertilizing, settling into a life of farming and animal husbandry that was to continue for centuries. The plow was invented when poking holes to plant proved too slow. The cutting blade and steel head were invented when the soil proved to be too full of roots or too thick for easy plowing. Genetics was developed when it was discovered that what was *really* needed was a method of increasing yield per acre. First the McCormick reaper came along, then the tractor, the combine, chemical fertilizers, hybrid varieties of grains, and on and on and on. In other words, the fact that Malthus missed was that human beings, unlike bacteria, elephants, or virtually any other animal on the face of the earth, when faced with a deteriorating environment, merely change the environment through the development of technology, and go on as before. As Figure 5-3 demonstrates, through this process of innovation, the top of the Malthusian curve is not removed, it is just pushed further and further back from the present. This brings us to the principle involved in much of real-world phenomena.

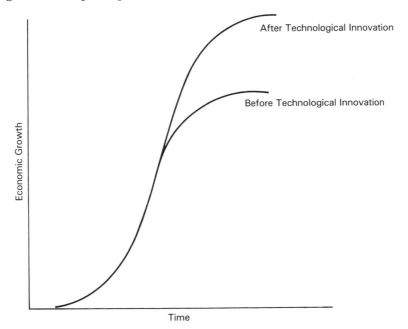

Figure 5-3 The Effect of Technological Innovation on the Malthusian Curve

SINUSOIDAL CURVES: THE GENERAL CASE

Malthus had discovered a condition that reflected a special case of a general class of real-world phenomena that describe how things grow through time. This was not limited to the growth of living things. Whether speaking of the growth of the human population or the change in the number of sunspots over an eleven-year cycle, the shape of the cycle is the same. It approximates a sine curve. Business cycles show the same effect, as do sound waves, ocean waves, electrical currents, the light curves of variable stars, the fluctuations of the Dow Jones industrial average, or the sleep cycles of mammals. In each case, there is a cyclical behavior pattern taking place through time, and that pattern tends to be sinusoidal in nature. Indeed, by definition, if it is cyclical, it must be sinusoidal in nature. The name given to this general description of growth through time is the *biological growth curve.* But what does this general case have to do with technology? As it turns out, the same phenomenon occurs in the technological cycles of modern society.

One such case has already been discussed, at least concerning the lower part of the curve. In the last chapter, the discussion on oligopolies referred to economies of scale as the cause of high market concentration. As industries grow, there is a tendency, particularly where technological innovation is an available means of increasing efficiency, to move toward bigness, thus reducing the number of companies in the market and increasing the market share that each of the surviving companies commands. It is in the economies of scale that we see reflected the lower section of the sinusoidal growth curve.

As technology is used to increase efficiency, the production rate per unit of input tends to rise, and rise faster and faster, thus indicating a geometric growth rate. This is identical to the general shape of the lower lobe of the sinusoidal curve discussed here.

The shape of the upper half of the curve is indicative of what takes place if this movement toward bigness continues. As a company expands and garners increasing levels of efficiency, it approaches a level of output in which the output resulting from additional units of input no longer indicate economies of scale. Beyond a certain point, the returns level off, and grow at a constant rate, as is seen through the center of the sinusoidal curve, where the rate of change becomes almost constant. This is the relatively straight section of the curve. Companies still increase their inputs since the outputs received are still valuable additions to their revenues, yet each unit of input seems to create the same relative output. The returns are constant, and this central area of the curve is said to indicate *constant returns to scale.*

The top portion of the curve represents the limiting case of technology's power to improve efficiency. If a company continues to increase inputs past the area of constant returns to scale, the output received from each additional unit of input *actually starts to decrease*, each unit of input being *less* efficient than the previous unit. Thus we see a characteristic reduction in the angle of increase along the top of the curve. Technically, the phenomenon is known as the *law of diminishing returns.* The law of diminishing returns states that after some point, the increase in output resulting from additional units of input will decline, first relatively, and then absolutely. In other words, after a certain point, each additional unit of input is less efficient than the last, and whereas there are gains from employing that last unit of input, the gain will be smaller for each new unit of input employed. This process continues until the result of adding one more unit of input is an absolute decrease in output. There is, therefore, a *maximum level of efficiency of inputs beyond which a system cannot improve, other things being equal.*

This maximum level determines the maximum size of a productive unit. An individual company or an individual plant owned by a company will have some maximum level of output of which it is capable. Beyond that level, further efforts are wasted.

If all three of these factors are joined, coupling the economies of scale with the constant returns to scale experienced through the center of the growth rate and the decreasing returns to scale that exist due to the law of diminishing returns, the result is a sinusoidal curve.

As with Malthus's population problem, the way to push the top of the curve for an industrial enterprise further away from the present is through the employment of new technology. A *given* technology has a maximum efficiency and a threshold beyond which it is unable to move. However, the introduction of a fundamentally *new* technology can push that limit ever outward. Thus power methodologies have their limitations, but as we move from foot power to horsepower to steam to internal combustion and electrical generation to fission to fusion and beyond, each new technology increases the available power for our use. The same is true of other technologies as well. There is a limit on the height of buildings that can be produced using wood and brick alone. Adding stone as a material pushes this limit further, but greatly increases the mass displacement of the building without a significant increase in usable space. By switching to steel-reinforced concrete, the mass–space relation is greatly enhanced. By shifting to radically new construction methodology such as the geodesic-dome concept, the ratio is further improved, allowing for monumentally large enclosed spaces at a relatively small outlay of physical resources. In each case, as one technology reached to the edge of its limits, approaching or entering the area of diminishing returns, it became more and more economically feasible to search out and develop new methodologies to overcome the problem.

Other examples are all around us. Compare the DC motors in size and output with the smaller, lighter, and more powerful AC motors developed by Tesla. Compare the load capacity and span capacity of simple bridge construction with that of the box-girder railroad bridge, or that of the huge expansion bridges, such as the Bay Bridge and Golden Gate Bridge in San Francisco and the first great expansion bridge, the Brooklyn Bridge, still a wonder after one hundred years of constant use.

So we see that technology represents a method of overcoming the sinusoidal nature of growth curves, representing a way of bypassing the hazards of the law of diminishing returns.

THE EXPLOITATION AND ACCEPTANCE OF TECHNOLOGY

In Chapter 2, the subject of homeostasis was explored, including the tendency of Homo sapiens to reserve judgment on the usefulness of technology as a survival mechanism designed to forestall a headlong charge into unknown new methods of operation that could be disastrous. Continuing this theme, the rate at which technology is accepted by a society and exploited to its fullest can both be expressed as sinusoidal curves.

Consider the life of the average product in the marketplace. No matter what the product may be or, for that matter, whether it is an unsuccessful, moderately successful, or highly successful product, it will follow a characteristic life cycle from introduction to discontinuance. This cycle is sinusoidal in nature.

At first introduction, a product is presented with much fanfare, and an advertising campaign is designed to put the name and image of the product before the public as often and as well as possible. This creates an awareness of the product in the minds of potential consumers. During this period, most customers are reticent to try the new product, being locked into their customary modes of purchasing by habit, homeostasis, satisfaction with present forms of the product in question, and a real desire to avoid the time-consuming process of seeking out and testing the new product to determine its suitability. In other words, the new product exists; consumers know it exists; but they do not feel it is sufficiently valuable to spend their money, time, and effort in its procurement and consumption.

There are some people, however, who are ready to try out a new product and are willing to go through the necessary processes to do so. They are called *innovators* in the marketing world. And the name is apt. These are people who are willing to innovate, to experiment, to try new ideas, and to enjoy the process. Their level of fear and homeostatic resistance to new ideas is lower than the norm, and they are more open to new concepts and new products. It should not be surprising that very often they are younger people, who are not so solidly locked into the characteristic cultural norms of their times, having not fully tested themselves in the maturing process.

Innovators will buy the product because it is new, because it is useful (or perceived to be so), or because it is different. This latter is the old "be-the-first-on-your-block" approach to individuality.

Once the innovators have begun buying the product, it will either be found to be useful by these pioneering consumers or found to be worthless. In the latter case, they stop purchasing the product and it falls by the wayside. If they like it, however, they will continue to purchase it, and the product will survive long enough to establish itself in the marketplace.

Now a second set of consumers discovers the product. These are called the *early majority*, and they represent a large number of people who are willing to try something new, provided it has been accepted by a sector of the population. Rather than rush right out to purchase the good, they will wait a while and see if it is still around next week or next month. They are the "wait-and-see" consumers who are willing to take a chance, if that chance does not appear too great. During this period of acceptance, the sales of the product will grow rapidly, will increase at a fast rate, and the product will become even more widely accepted and seen in society.

At this point, the final holdouts, the *great majority*, will "discover" the product, as it is now seen as useful, has been tested by the more progressive of the population, and has become an accepted part of the culture. Once this part of the consuming population accepts the product, its sales start to level off and it tends to peak at what is called *maturity*. From here on, it is a matter of finding ways to maintain the high sales levels of the mature product and to keep customers from being lured away from this product to try others. Eventually, the product will decline as new products take its place in the market, and consumers move on to other, more interesting ways of creating satisfaction through consumption.

If this process sounds familiar, it is probably because it describes a sinusoidal growth curve. Technology, to be accepted in the marketplace as a means of accomplishing ends, must go through the same acceptance phases as any other product or method that is being introduced. An early majority will always be willing to try it, but unless they decide that the new technology is useful, it will be rejected. If the innovators continue to use it, others will consider it, and the early majority begins to discover it as a means of accomplishing ends as well. Finally, it establishes itself as a benevolent methodology and is generally accepted, being used by much of the society for the good of the society. Innovation is no less a product than any other good and can be viewed as such. The questions to consider are how quickly each of the stages will be spanned and how long it will be before the technology is accepted.

And who is to say that a given technology must be accepted at all? There are many examples of technologies that were perfectly viable and yet were rejected. Their rejection stemmed not from their lack of viability but from the lack of need for what they were capable of doing. Technologies are only used if they are useful. They must wait until their time has come or be passed by.

TECHNOLOGY WHOSE TIME HAS COME

Technology, like any other social element, will only exist as long as it is useful to the ends of the society. Just as the various economic systems disappeared as conditions changed, so will a given technology give way to another when the old is no longer capable of accomplishing the goals of the society in the face of current conditions.

And yet, the time between the development of a technology and its widespread use by a society may considerably vary from instance to instance. In the case of flying, the acceptance of the technology was rapid, being spurred on by the visionary actions of a few innovators who built on the original work of early pioneers such as Claudius Dornier, the Wright brothers, and Glenn Curtiss, and by the added impetus of World War I. The world moved from Kitty Hawk to space in less than fifty years! On the other hand, the steam engine had a very long initial acceptance process. First developed in 1687 by Denis Papin, though not existent as a working model until Thomas Savery's version was finished in 1698, the steam engine was not in common use until after the first model of James Watt's engine, patented in 1769. It was not available for sale until 1775. Why the difference? What creates the delay in one form of technology while another takes nearly a century to even begin to receive acceptance? Two examples should suffice to indicate the difference.

Consider a technology that is really a method more than a technology, that of Boolean algebra. Anyone who has ever dealt with computers beyond a very superficial level knows well the importance of the Boolean algebraic methods in the way a computer works. The Boolean system is binary, having only two numbers in it, 0 and 1. It is through Boolean algebra that these numbers can be manipulated to represent values as powers of two, and thus allow large numbers to be stored in electronic switches as either on (1) or off (0). Yet the methodology is nothing new.

Boolean algebra was developed in 1854 by an English mathematician, George Boole. It is alleged that part of his motivation in creating the algebra was to win a prize in mathematics being offered for the most original new concept in the field. Boole merely noted that deductive logic was a system that could be symbolically described using the language of mathematics and that, if certain rules of manipulation were formulated, these mathematical symbols could be utilized in combination to come to deductively logical conclusions. This is essentially what Boolean algebra does, using 0 and 1 for true or false and utilizing them to build truth tables for the solution of complicated deductively logical problems. This is all very well, but unless you happen to be a deductive logician, it may appear to be of little relative value.

Or is it? In light of what we now know, obviously not. Claude Shannon, an engineering student, noted in his master's thesis that the Boolean algebraic format could be applied to logical problems using the binary system. This was in 1938, some eighty-four years later! What was at first a mathematical oddity, of use only to those interested in the more esoteric forms of logical construct, has become the basis of an entire revolution in science, technology, and, indeed, culture.

It was not until the time was right for the fruitful application of the method that it gained importance. Thus, when it was first discovered, it was largely ignored because of a *lack of application*. Once the applications arose, the explosive use that it experienced was incredible. It is now in the midst of an ever-increasing growth pattern, and is not yet fully utilized by the society.

A second case shows the exact opposite pattern. The laser, a device for focusing coherent beams of visible light (laser, as you are probably aware, stands for *l*ight *a*mplification by *s*timulated *e*mission of *r*adiation), was first proposed in 1958 by Charles Townes, who shared the 1964 Nobel Prize in physics for his work in developing a similar procedure for microwaves, the *maser*. By 1960, two successful working models had been developed, one by Theodore Maiman and a second toward the end of the year by Ali Javan, at Bell Labs. Almost immediately, physicists and electronic engineers were investigating the possibilities represented by this wondrous new device. Because of the special properties of lased light, the applications were staggering. It created a highly accurate, thin, closely focused beam of light of considerable power that could be used in many fields. The laser became known as the "solution in search of a problem." In just twenty-five years the laser has been utilized in weapons and weapons guidance, eye and cancer surgery, communications, measurement, fusion research, information retrieval and storage, holography, photomicrography, electroholography, chemical research, nuclear physics, and many more.

In this case, the conditions encountered by the Boolean algebraic process were not present with the invention of the laser. The laser came about at a time when there was a crying need for it. Investigators were on the edge of new discoveries in a number of sciences but were unable to continue because of a lack of technology. The laser provided that new technology, and expansion took place in many fields. The time was ripe for laser technology, and, as with the Boolean process, we have only just begun to find uses for it.

Yet this "right time and right place" element is not alone in creating the growth curve of a technology. Also to be considered is the matter of awareness. Information expands geometrically if it is made available to a population. One person tells two people, who each tell two others, and so on, the rate of dissemination growing geometrically. And with this comes an expansion of the application of a technology. Early on, when the population becomes aware of the existence of a new technology, it applies that technology to all of the obvious problems for which it is a solution. And as more and more people become aware of its usefulness, that application expands rapidly, as with the laser. Then, as the obvious applications begin to be utilized fully, the less obvious ones are allowed a chance. Since they are less obvious, it takes somewhat longer to realize their potential. This is the steadying growth rate experienced in the sinusoidal curve form after the initial explosive flash of application, as with the laser. Finally, when the technology is mature, it reaches the limit of its application, and the expansion of its use slows, turning the sinusoidal curve at a slower and slower rate, until it peaks. At this point, the technology is fully utilized and has probably become an integral part of the culture.

The steam engine was initially utilized for a host of static applications, such as pumping water and running mills. Later it was successfully developed into a transportation power source with the advent of the locomotive, and, in the early twentieth century, the steam automobile. In addition, it found application in the production of electrical power, first at the hands of Edison, who opened his steam-generated electrical power plant in New York City at the Pearl Street power station in 1882, and later in municipalities across the nation where hydroelectric power generation was not feasible. In the production of power for ships it improved through the paddle wheels of the early nineteenth century to a zenith of efficiency with the steam turbine developed by Charles Parsons in 1894 and adopted as the standard propulsion of the British Navy shortly thereafter.

The nineteenth century was the age of steam just as the twentieth has been the electrical age (no matter how often we try to convince ourselves that it is the nuclear age, an epithet that is not yet logically applicable). Steam still exists as a motive force in our society, but its useful life as a technology is in the final stages of maturity, having given way to more exotic, modern, and appropriate forms of power production.

Technologies run a course from infancy through maturity in a manner similar to the life of a given paradigm which slowly replaces earlier, less useful paradigms and is finally replaced by a new one once the solutions the paradigm offers are exhausted. Each of these instances is reflected by a sinusoidal curve. As such, the characteristics of a technology and the way in which it interfaces with the rest of the social system can be understood by determining its position on this continuum.

CONCLUSION

The biological growth curve format, the sinusoidal curve, describes a wide range of real-world phenomena, particularly phenomena that deal with development or growth through time. It is equally applicable to biological and nonbiological examples. Also, by studying technology in terms of growth, maturation, and decline, it becomes possible to increase our ability to understand what has taken place, what is taking place, and, most importantly perhaps, what will probably take place in the future. The characteristic slow startup of technology in its developmental infancy can be described as the beginning of the sinusoidal growth curve, with homeostatic, communication, and application problems being the main causes. The following rapid adoption and expansion of application can be seen as the increasing growth rate of an expanding system, followed by the steady growth of a fully internalized technology, being developed to its fullest. As the utilization of the technology reaches its zenith, the high end diminishing returns will create a slowing of growth to the point where the technology is fully utilized. It peaks here and either continues as a useful methodology or declines as new technologies replace it as a means of solving new and extended problems for the culture.

THOUGHT AND PROCESS

1. If you have access to a petri dish or other flat glass dish, try to duplicate the biological growth curve. In the plate, place a solution of nutrient material (pure gelatin works well) or a small piece of moist bread and periodically observe the growth in colonies that develop. For an added dimension to the experiment, prepare two plates. Leave one open to the air to allow a continuous inflow of microorganisms, and cover the other with a jar or some other transparent cover to limit further influxes of migrating spores after the first initial colonies appear. In this way, it is possible to study by analogy the effects of immigration on the life cycle of an environment, as in the difference between population growth of North America before and after the migration of Europeans began in the 1500s.
2. The development of applications for technology has been purported in this chapter to follow some type of sinusoidal curve form. To test the truth of this, try a simple brain-storming technique. With four or five others, pick a technology from the following list of hypothetical technologies and speculate on the

possible uses to which the new technology could be put. Have one person write down the ideas in brief, while an outside party records the *rate* at which the ideas flow through time. (An easy way to do this is to note the time passage periodically by starring the last idea listed each two minutes, say, in ten-second intervals.) Give yourself about fifteen minutes for this exercise to be sure that all possibilities are explored. Also, allow yourself to loosen up enough to verbalize some of the more far-fetched applications that you think of. The results of this experiment, when plotted, should approximate a biological growth rate curve. *List:* (a) an antigravity machine, (b) a teleportation device, (c) a faster-than-light engine, (d) chemical intelligence boosters, (e) a means of tapping the atmosphere for free energy, (f) telepathic communications, (g) a worldwide master computer, and (h) direct interface between the human brain and microchips.

3. With a little research (in some cases, none) it is possible to get some sense of where we stand in relation to our technology. Draw a sinusoidal curve, and place the following list of common technological devices and methods on the curve in terms of their position in their useful life. Disregard whether the individual technology has been in existence for a long or short time. Concentrate on their position on the curve, not on their relation to each other. *List:* (a) the laser, (b) the steam engine, (c) electricity, (d) automobiles, (e) diesel engines, (f) the computer, (g) the internal combustion engine, (h) genetic engineering, (i) steel, (j) plastics, (k) wood, (l) refrigeration pump technology, (m) airplanes, (n) glass, (o) electronic synthesizers, (p) television communications.

4. Research the discovery and development of Teflon as a product. How long did it initially take the innovators to discover the product? How long for the upswing in use by the early majority? Where are we now on the Teflon-use curve? Why?

5. One of the most rapid discovery-adoption-maturity-decline patterns available in the real world is the fad cycle of certain products. Among the products exhibiting this rapid rise and decline are such things as hoola-hoops; yo-yo's (which seem to return every generation or so); pet rocks; and knickknacks with certain themes, such as frogs, cats, mushrooms, rainbows, and cartoon characters. The skateboard is an example of a fad that never quite died. In light of our discussion in this chapter, what other fads can you think of that did not die, and what is it that makes them different from the ones that did? How does this reflect on the acceptance and maturation cycle of technology?

Television and Cultural Change

By
Fred Aboundader
and
James A. Hutchinson*

INTRODUCTION

Is television one of the most beneficial inventions ever devised by humankind? Or is it one of the greatest curses? In this report we are going to examine some of the pros and cons of TV and let you, the audience, decide. We have divided our subject into six categories, as pointed out here in our outline.

1. Development of television
2. Influence of television
3. Corruption of television
4. Educational aspects of TV
5. Communication capabilities of TV
6. Future aspects of TV

DEVELOPMENT OF TELEVISION

Modern television, now viewed by millions of people around the world, is a result of over eighty years of research and development. Scientists in Britain, Germany, France, the USSR, and the United States contrib-

*Fred Aboundader and James A. Hutchinson, "Television and Cultural Change" (an unpublished paper presented at DeVry Institute of Technology, Atlanta, Georgia, April 24, 1984).

uted to early experiments in television. However, it was Britain, the United States, and Japan that solved the problems leading to a full television service.

The dream of extending human vision, as the telephone had extended human voice, began to be realized in 1883. In that year, a German scientist, Paul Nipkon, invented a scanning device that would break down an image into a sequence of tiny pictorial elements. This crude mechanical scanner was used with a photoelectric cell that converted light into electrical impulses.

Experiments with mechanical scanning were pursued in the 1920s. In 1925 Charles Francis Jenkins, an American inventor, used elaborations of the Nipkon disk to broadcast silhouette pictures from his workshop in Washington, D.C. The Scottish inventor John Logie Baird made a public demonstration of television in 1926, but he only produced shadow pictures. However, by 1928 he had broadcast television pictures in color, outdoor scenes, and stereoscopic scenes.

Ernest J.W. Alexanderson began daily TV tests on the experimental station W2XAD in 1928. On September 11, 1928, General Electric (GE) presented the first dramatic production on television. It was "The Queen's Messenger," with the sound carried on the AM radio station WGY. In 1931 the Radio Corporation of America (RCA) made experimental tests over station 2XBS in New York. David L. Sarnoff, president of RCA, predicted that within five years television would become as much a part of our life as radio.

Despite these early successes, mechanical scanning had inherent drawbacks. In particular, it did not provide sharpness of detail. Consequently, further advances depended on the development of electronic scanning.

The idea of electronic scanning dates as far back as experiments by Heinrich Hertz in the 1800s, to the publication of the theory of photoelectric effects in 1905, by Albert Einstein, and to Karl Braun's discovery that he could change the course of electrons in a cathode-ray tube. It was in 1907 that the English scientist Alan Cambell used a cathode-ray tube at the receiving end.

World War I put an end to all but theoretical work, but its aftermath brought to America one of the outstanding scientists in the development of modern television. He was Vladimir K. Fworykin, and in 1923 he had developed a crude, but workable, partly electronic TV system. In 1930 Thilo T. Farnsworth developed a new electronic television system that made TV pictures suitable for the home. In 1939 the first television sets were manufactured for the public.

It was in 1935 that David Sarnoff, president of RCA, invested $1 million in television program demonstrations. Also in 1935, the British

Broadcasting Corporation (BBC) took over the control of British television from experimenting companies.

It was in 1939 that the National Broadcasting Company (NBC), a subsidiary of RCA, began a regular TV service. 1939 was the first time a U.S. president was televised, the first broadcast of a major league baseball game, and the first broadcast of a college football game.

In 1941 frequency modulation (FM) broadcasting was adopted for transmission of sound for television. In 1941 the Federal Communications Commission (FCC) authorized commercial TV beginning on July 1. WNBT, New York, became the first commercial station. CBS presented the first television newscast on December 7, 1941, reporting the events at Pearl Harbor. Consequently, World War II greatly slowed down the development of TV and what television broadcasting remained was used for purposes of civil defense and air raid warden training, Red Cross instruction, and U.S. bond sales.

By 1948 television was well on its way. There were thirty-six television stations on the air, seventy under construction, and about one million sets in use. However, on September 30, 1948, the FCC declared a freeze on the licenses of any new TV stations in order to study frequency allocations. When the FCC lifted the freeze in April 1952, it allocated more than 500 stations to the VHF band and more than 1400 stations in the UHF band.

Now we have more than 150 million households with television, and the rest is history.

INFLUENCE OF TELEVISION

Indeed, it can safely be said that no other force, in so short a time, has exerted such a powerful influence on so many people. In the last three decades it has had an enormous effect on family relationships, entertainment, education, politics, advertising, news, sports, and other areas of human endeavor. Such cases reflect TV's increasingly pervasive influence on America, both good and bad. In America, where television has become the eyes and ears on the rest of the world, TV is a primary force determining how people work, relax, and behave. Recent studies show that the lives of Americans from their selection of food to their choice of political leaders are deeply affected by TV, and the influence is growing.

Let's take an example here: When one woman's TV set broke down, she said, "It's like somebody in the family just died."[1] Some

---

[1] Frederick W. Franz, "What Are People Saying About Television?" *Awake*, April 22, 1982, p. 3.

people admit to being "TV-intoxicated," needing a daily fix of it as a drug addict would need drugs. That is an interesting point, because a lot of people are TV addicts and don't even know it. How can you tell? Ask yourself these few questions, and see how you answer them.

1. Do you look forward to the end of the day so that you can watch your favorite TV program?
2. Do you keep the TV on after your favorite program is over and keep watching others?
3. Do you do the above (1) and (2) night after night?
4. Do you keep the set on even when you are not actually watching it?
5. Do you turn the TV on in the morning if you have the opportunity?
6. Are you irritable during an evening if you cannot watch TV?
7. Do you make excuses for watching too much?
8. Do you spend more hours watching TV than all other leisure activities combined?
9. Would you rather watch TV than be with friends or do things with the family?
10. Do you become defensive if accused of watching too much TV?[2]

If you have answered yes to a number of these questions, then this suggests that some degree of TV addiction has already set in. Don't fret. Not only do I have questions, but I have some possible solutions. To help control TV watching, some people have put their set in a place where it is inconvenient to spend long hours with it. Some have put the TV in a cabinet or a closet, requiring effort to prepare for viewing. Also, since a bedroom is too conducive to lying down and watching for long periods, many will not have a TV there. Research shows that older women, 55 and over, watch the TV the most on a weekly basis. This seems possible because older women generally outlive men, and many may be poor and may not be able to afford other forms of information and entertainment. Also, the elderly and shut-ins have found TV useful since it helps to combat loneliness.

Another example of powerful influence is the advertisers who use TV as a medium of communication. Their advertising conditions people mentally, so that they will buy their products. Advertisers invested some $11 billion in 1980 in order to present their commercial messages before the public.

[2] Franz, p. 17.

Another point discovered in doing this report was how TV affects family relationships. Some publications state that families are brought together by television. This may be true from a physical standpoint, but we believe it has actually created a communication gap at home. In essence, families do not get the opportunity to share day-to-day experiences or participate in the normal give and take of family communication due to the competition of prime-time TV viewing. TV makes it harder for some people to relate to others because it is difficult to make the transition from watching TV to real people. This may be because viewing TV requires no effort, whereas dealing with real people requires effort.

Another harmful effect of too much TV viewing is in regard to actual physical health. Many people eat between meals while watching TV, which contributes to excessive weight problems. Television viewing requires no physical participation. Prolonged periods of inactivity are detrimental to the human body; moreover, inactivity can also be detrimental to the human mind.

CORRUPTION OF TELEVISION

Television with all its pluses in communication and information has one big minus—its corruption. I am not talking about the corruption of the television industry, but rather the corruption of American society. This corruption of society is brought about by the frequent use of crime, violence, sex, and abusive language on the home screen. This continual viewing of violence, crime, and sex affects all ages, but none more than children and teens.

Children, the most easily influenced creatures on the earth, view TV on an average of about 4.2 hours daily; this is approximately 33 percent of total daily awareness hours. It is true that most children, ages 6 to 13, spend an average of 3 more hours per day viewing television than studying. This would not be so bad if the television shows they watched were informative and educational. Since 1975 public complaints about violence and sexual items on programs have rapidly increased.

For instance, in Los Angeles in January 1975, Metromedia-owned KTTV dropped a one-half-hour segment of the Superman, Batman, Aquaman cartoon. It was labeled violent for children by the local chapter of the National Association for Better Broadcasting (NABB). Also in 1975, a California woman sued television producers of the film *Born Innocent*, charging the program's graphic rape scene may have stimulated a similar sexual attack on her daughter. It isn't that the public isn't aware of the violence, sex, and abusive language; on the contrary, they (the public) have accepted it as "normal" in their society.

Children cannot discriminate between reality or fiction in television shows. What they view is stored subconsciously in the back of their brain to be recalled later on in life. Studies show that children who have viewed violent TV productions at very young ages are more disruptive and violent than their peers who viewed mostly educational shows, such as Sesame Street or Mr. Rogers, or those who viewed very little television.

Every day on cable television's MTV, children as young as five regularly watch women in chains and people being tortured and shot.[3] This was stated in the *Atlanta Journal and Constitution*, April 16, 1984, afternoon edition. What is so ironic is that on December 20, 1982, Jabari Siniama, access director for Cable Atlanta, was quoted as stating, "In no other city in the country will you see anything like it."[4] What he was referring to was a video on New York City cable that showed a rock group chainsawing a female victim to death. Now, less than two years from when he was quoted saying this, Atlanta has its own Atlanta Rock Review, which airs videos often viewed as equally distasteful as the one in question in New York City.

In the May 1982 issue of *Gallup Report*, a monthly magazine that reviews statistics and surveys, the cover article dealt with TV and crime. The report sees a cause-and-effect relationship between TV and real-life violence. As television has increased its broadcasting of crime and violence on public cable, real-life crimes, especially sexually related and violent crimes, have increased. For example, in 1976 Los Angeles police asked NBC to set up a special screening of a "Police Story" episode that some believe may have inspired a killer to slash the throats of three sleeping derelicts on skid row. The plot of the "Police Story" program featured the same kind of crime.

The Federal Communications Commission also noted that nationwide levels of crime and violence are rising; it observed that an inordinate number of high-tension, crime-drama shows are running night after night on prime-time television stations across the country. So why do networks continue to show explicit violence and sex despite the adverse results?

Network officials assert that most programming is determined by extensive survey of the public. "Basically, networks produce the kinds

[3] Barbara Jaeger, "Violence, Rock Music, Counterculture Art Could Pose Problems for Media Culture Kids," *Atlanta Journal and Constitution*, 16 April 1984, 1-B.

[4] James Gosh, "TV Video and Children," *Wall Street Journal*, vol. 20, December 1982.

of programs that the audience demonstrates it likes," said Herb Jacobs, president of Telkom Associates. "If the public wants police stories and Westerns, it's pretty hard to dramatize them without violence."[5]

The most obvious change is in the language being used on the air. Only a few years ago words such as ethnic characterizations would not have been broadcast. Evidently, today these slang terms are used commonly in everyday dialogue.

All these changes of more violence, crime, sex, and abusive language on prime-time television reflect changing attitudes in American society. Basically, networks give the public "what the public wants."

EDUCATIONAL ASPECTS OF TV

One beneficial use of TV is for education; good TV programs can certainly teach children and adults many things. From the beginning, educational television has had a dual purpose: to supply programs for school use as a supplement to the teacher during the day and to supply programs of community interest and cultural enrichment at night. Early studies showed that teaching by television generally was more effective than most classroom procedures; therefore, educational television networks are very promising. As for cultural programming, the need for it increased as commercial TV became less and less a service and more and more a business. Since 1952, the number of channels reserved for educational TV has increased from 242 to 309.

As a matter of fact, some local colleges use TV stations for teaching credited college courses. Dekalb College, for instance, teaches such subjects as sociology, psychology, and computer programming by TV. The students only come in four times a quarter to take examinations. I don't know how effective this system is, but it seems to be freeing the student from actual class time.

Experts say that, when used properly, TV can stimulate reading. It can present ideas that encourage viewers to want more information so they will get reading materials that will add to their knowledge of the subject.

I want to deviate from the subject of education to tie in some other significant point. Not only does good programming teach the public but bad programming does an equally effective job. For example, convicts have admitted to getting ideas for crimes by watching programs in prison. In one survey, a surprising 90 percent said that they had actually learned ways to improve their criminal techniques through

[5] John Anderson, "What Is TV Doing to America?" *U.S. News and World Report*, August 1982, p. 26.

TV viewing. Four out of ten said that they had already tried specific crimes that they first saw on television. One prisoner said, "Television has taught me how to steal cars, how to break into establishments, how to go about robbing people, and even how to roll a drunk. Everybody is picking up on what's on the TV." [6]

COMMUNICATION CAPABILITIES OF TV

Since World War II, television has become the most popular mass communications medium in history—perhaps the first mass medium to reach all segments and groups in a society. Next to sleeping and working, Americans spend the greatest part of their time watching television, and TV viewing clearly is a pervasive social activity in some other countries as well.

TV is an effective medium of communication because it brings information about current events to us faster than do magazines or newspapers. And it does so in a form that is highly interesting to the eye: in motion pictures. We feel as though we are actually present, witnessing what is going on sometimes thousands of miles away.

The transmission of commercial television across an ocean first became possible in April 1965, when Early Bird (Intelsat 1), the first commercial communications satellite, was placed over the Atlantic. International TV broadcasting on a global basis was first established in mid-1969 by a global system of Intelsat 3 satellites.

In addition to transmitting television, telephone, telegraph, digital data, and facsimile signals simultaneously, the Intelsat 3 satellite has a multipoint communications capability that is particularly useful for distributing TV programs. What this means is that programs produced in the United States could be broadcast for simultaneous viewing in countries all over the world.

In researching further, TV in the 1960s provided a number of other firsts in the field of communications. For example, TV became a triumph of communications technology, highly interesting, and historical on that memorable day in July 1969. Of course, many will recall that when Neil Armstrong and Edwin Aldrin made the first manned landing on the moon's surface, an estimated 600 million people in some fifty countries on six continents witnessed this event as it was happening.

Also, TV had another first in the area of politics when it was used to air the debate between John F. Kennedy and Richard M. Nixon. These confrontations gave the candidates an opportunity to discuss differences on issues for immediate rebuttal. This gave the public a chance to see and hear the candidates. An interesting comment made

[6] Franz, p. 10.

by the Mass Communications Research Center regarding the results of the debates was, "Kennedy did not necessarily win them, but Nixon lost them."[7] The debates definitely did much to create John Kennedy's "image."

After Kennedy won the election, he was credited with introducing "live" TV press conferences from the White House.

Some people have grown increasingly troubled by some of the effects of TV on democratic government. In 1980 networks declared Ronald Reagan the projected winner soon after polls in Eastern states closed but before balloting ended in the West. Experts say some prospective voters never went to the polls in the West, believing their votes would make no difference.

To sum this up, we can see that the TV, as a medium of communication, is the most effective means of mass communication known to humankind. It has linked together over 118 countries—all nations in the Western Hemisphere, all nations in Europe, over one-half of Africa, and most of the nations of Asia. These countries all have a privately, publicly, or governmentally owned television service.

FUTURE ASPECTS OF TV

Television is a very important aspect of today's society, and it will be much more of tomorrow's. With technology spreading so rapidly, we might all be wired up to a two-way television system as early as the year 1990.

A two-way television system is where we, the viewers, have a computer tied to our cable system and are able to give the station some kind of feedback. Already in America there are over 90,000 homes with some type of two-way system; however, a large number of these systems just have response buttons and not computers. Two-way systems allow the viewers to respond to opinion polls, advertisements, and shows; they can also allow subscribers to do their shopping and banking from the convenience of their homes.

For example, an average housewife's morning might consist of going to the grocery store, picking up a few items at the department store, paying the utility bills, and depositing a check at the bank on the way home. All this might take several hours.

However, if she were living in a "wired society," she would only need to walk over to her television set and spend maybe 15 to 20 minutes typing in her morning chores. This leaves her 4.5 hours to do other things.

It may seem now that the telephone system has the market on two-way communication, but it may not be so six years from now. It

---

[7] "Television," *Encyclopaedia Britannica: Macropaedia* (1981).

is already possible to hook the television into the telephone and have vision as well as voice. TV-phones have been around for over two decades, but weren't feasible until now with the invention and application of the microprocessor.

Imagine everybody with a TV-phone: your friends, your family, the hospital, the department stores, or maybe your date for the evening. This would almost eliminate so-called "blind dates." TV-phones will bring people who are far apart very close together. They will see each other's clothes, hairstyles, and facial expressions. As the old saying goes, "A picture is worth a thousand words."

Another aspect of television in the future is the screen size. Television screens will get larger and larger; this is already true of the new projection television.

When you think of a large-screen television you might think of a 21-inch diagonal, but the screen size I am referring to is about a 10-foot diagonal. You might think I'm crazy, but think about it. With the rapid advancements in digital techniques, it may not be too far away.

Also I believe that as technology advances and the screens get bigger, the picture itself will become more realistic. What I mean is that the television will broadcast more three-dimensional (3-D) movies on the air, but the 3-D I am referring to will be more realistic than in the past. Television broadcasts in 3-D combined with television sets capable of receiving 3-D broadcasts will give a much better portrayal of reality.

Overall, however, the most important part television will play in the future is as a two-way communication medium. Once the lines have been installed and the computer-based two-way systems are more common, there will be no end to its limits.

BIBLIOGRAPHY

Anderson, John, "What Is TV Doing to America?" *U.S. News and World Report*, August 1982, pp. 24–29.

Franz, Frederick W., "What Are People Saying About Television?" *Awake*, 22 April 1978, pp. 3–20.

Gosh, James, "TV Video and Children," *Wall Street Journal*, December 20, 1982.

McLuhan, Marshall, *Understanding Media*. New York: The New American Library, 1964.

Naisbitt, John, *Megatrends*. New York: Warner Book, 1984.

Peter, Fred, "Future World of Television," *Science Digest*, September 1980, pp. 16–21.

Salgado, Robert, "Cable TV Video Violence," *Atlanta Journal and Constitution*, 16 April 1984, 1-B.

"Television," *Encyclopaedia Britannica: Macropaedia* (1981).

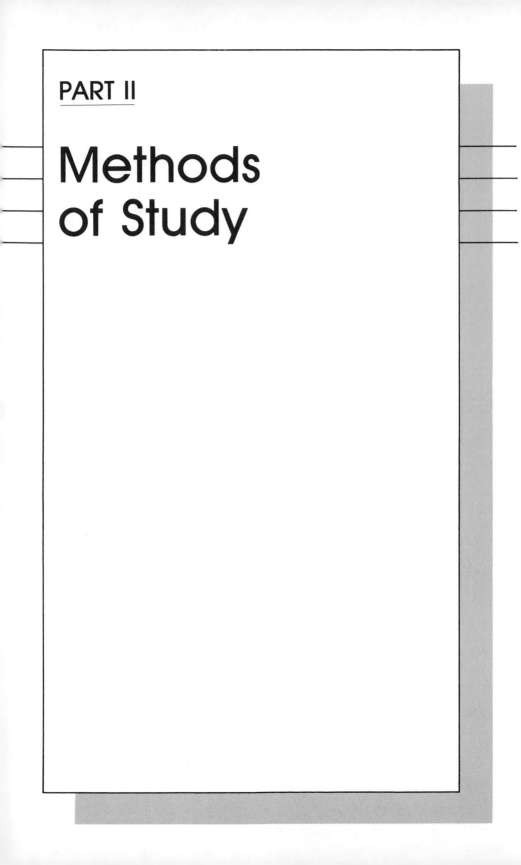

PART II

Methods of Study

CHAPTER 6

The Cause and Effect of Technology and Society

CAUSE AND EFFECT: THE ROOTS OF PREDICTION

If something occurs once, it is an incident. If it occurs twice in the same way and under the same circumstances, it is a coincidence. If it happens again a third time, then we have the beginning of a pattern, a way of proving cause and effect. The comparison of events to determine patterns is something that we have all experienced throughout our lives. If we miss a parking place whenever we arrive at work after 9:30, we learn that there is a pattern, and attempt to arrive prior to 9:30. If certain foods create discomfort whenever they are eaten, we learn not to eat them after a very few trial runs. If the new boss is grumpy three Mondays in a row, we expect a similar mood on the fourth. And in each case, our expectations about the future are based on a comparison of similar events from the past.

Cause-and-effect relationships are a fundamental element in studying the manner in which various aspects of society affect one another. It can be seen from the preceding chapters that the desires of individuals interacting in the economy caused their behavior so that a given result (effect) was created. Given the desire to maximize profits (the theoretical aim of any economically rational human being), the result was the production and distribution of goods and services in the economy. This is purely a matter of cause and effect.

And the fact that certain actions were taken in order to create the desired results is an indication of how much people depend on their understanding of cause-and-effect relationships to create the world the way they wish it to be. All of us, every day, live in accordance with our understandings of the cause-and-effect relationships that exist in the physical world. As such, we are predisposed to view things in the light of this knowledge.

Imagine what it would be like if the world were constructed in a random manner, or, just as importantly, if people behaved as if the world were randomly ordered. What could we depend on? What could we know about the future? What would we expect to happen with our every move? It would be impossible to get up in the morning and go through a normal routine without predisposing ourselves to the belief that the cause-and-effect relationships that we had experienced in the past were going to hold true today as they had before. The sun has risen every day of your life. Without cause-and-effect assumptions, you would not know to expect it to happen again tomorrow morning. In fact, it is this determination of cause-and-effect patterning that allows us to study and manipulate the physical world in the first place.

We make these assumptions about what will and will not happen on the basis of statistics, which brings us back to the statement made at the beginning of this section: One event is an incident, two identical events are a coincidence, and three begin a pattern that could indicate a cause-and-effect relationship.

We are always gathering data and comparing them with other data, counting the times certain events coincide. On the basis of this analysis, we reach conclusions that are assumptions about the future. The more data we have supporting that two events are connected, the greater our belief that, given one set of causative factors, another set of results will occur. With three events we have sufficient data for a pattern. With less than three, there is not enough security in our understanding of the situation to allow any degree of confidence in our predictions of what will happen next. And we do this whether we are consciously aware of the process or not.

As we record more and more instances of circumstances and events, we increase the "odds" that our assumptions of cause and effect are correct and are more and more secure in behaving in accordance with those assumptions.

Consider the infant who is constantly going about the process of learning what works in its new world. Reaching out and experiencing objects creates a store of information upon which to base expectations. The infant is fascinated with everything. Anything new, anything different that is within reach and within the power of the infant to explore, it will probably investigate. Bright colors, interesting objects of new shape or size, the soft cooing of a parent's voice, or the sound of a music box are all eagerly explored by the infant, attempting to gather data. And when the same objects are experienced in the same way over and over again, the infant is able to create patterns, to connect characteristics that define objects as different from one another. They can be classified as fun or not fun, enjoyable or not enjoyable, new or previously experienced. They can be classified as something to be sought, such as a ball that makes a delightful tinkling sound when it rolls, or something to be avoided, such as a hot light bulb that causes pain when touched, or an object with which the baby is not allowed to play, resulting in rejection behavior from parents when the forbidden object is touched.

Many cause-and-effect relationships are learned in these first years. If you drop a ball, it falls; if you touch a hot light bulb, it burns. If you dribble your food, your patient mother or father merely puts more in your mouth until you learn to swallow it properly.

And we continue to make these associations through life. They are the pigeonholes into which we place our understandings of experiences and form our world view. As mentioned in the discussion of creativity, this paradigm construction can limit our ability to deal with problems by narrowing our focus. Yet without the cause-and-effect processing, we would be hard pressed to progress beyond a very primitive stage of existence.

RIPPLES IN A POOL

To the social scientist or to the technologist studying the effects of society and technology on each other, the importance of the cause-and-effect relationship lies not only in its basic nature but also in the fact that each effect becomes a cause and each cause is in reality the effect of some other cause. This is an integrated system, in which no single entity can act or react without causing further results (effects). The phenomenon is often referred to as the "ripples in a pool," analogous to the spreading wave patterns emanating from the locus of a pebble dropped into a body of water.

When a single event occurs, it causes other events to occur. Each of those events will have consequences, causing, in turn, other events, so that the occurrence of a single event spreads its influence in ever-widening patterns, affecting in some way every element of the system in which it exists. For most events, these effects are small, but for some, they are quite large, and, unfortunately, not always immediately apparent. This was the case cited in the example in Chapter 7 of the producer of farm goods who expelled wastes into a nearby stream. The widening ripples of the producer's actions encompassed the entire community within which the plant existed and, if followed far enough, will have some minuscule effect on the economy of the entire world. Primary causes create effects that result in further effects, called *secondary effects*, which further result in more effects, which are called *tertiary effects*, and so forth.

Following this process to its logical conclusion would seem to lead to the assumption that everything is affected by everything, and that may be the case. How then can we possibly study the effects of a single change in technology or the rise of a single sociological factor? How can such a massive number of cause-and-effect relationships be reasonably dealt with? For this, we must turn to the study of systems and systems analysis, which reduces huge interactive processes to manageable, organized forms.

SYSTEMS AND SYSTEMS ANALYSIS

The System

A *system* is an aggregate of two or more physical components and a set of disciplines or procedures by means of which they interact. The physical world is itself a system, and contains innumerable systems within it. This broad definition includes any and all groups of physical components that interact through a set of procedures. These procedures are definitions of the cause-and-effect relationships that exist in a system.

An automobile is a system. It is composed of a finite number of highly specialized physical components, each of which has a specific relationship to other components in the vehicle. A piston can only do certain things, and only does them in response to certain causes. When there is oxygen, gaseous fuel, and a spark present in the cylinder with which the piston is interacting, the resulting explosion causes the piston to move away from the cylinder head under pressure from the expanding gases. At its fullest recession, the crank shaft inhibits any further retreat in the receding direction, valves open to allow the gases to escape the chamber, and the crankshaft forces the cylinder to again move toward the cylinder head, where the same set of actions takes place again. Physical components—the gaseous fuel, spark plugs, oxygen, cylinder and cylinder head, piston and piston ring, crankshaft and connecting rod—all interact through certain laws (the laws of thermodynamics, the behavior of expanding gases, the mechanical laws of axles and cylinders, and so forth) to behave in the same way over and over again. This is a system.

It is also a subsystem. Each cylinder and cylinder system is only one of four (or six or eight) that all perform their functions in a coordinated manner to create the engine. And a number of other systems also function in this task. Taken together, we have a system known as an engine. In concert with the other components of the vehicle, the engine is part of a system known as a car, which is part of a system known as a transportation system, which is part of the social system, which is part of . . . .

Systems need not be mechanical in nature. They may be animate as well, or may be a combination of mechanical and animate, as in the case of integrated systems such as a space shuttle, where both people and machine are components of the whole. The Battle of the Bulge was an interaction of two systems. The voyage of Ferdinand Magellan was a system as well, as was the Transcontinental Railroad and the Standard Oil Trust. Major corporations are systems like the human body or the political system of France. These are all systems, each with a given purpose, made up of elements proper to the performance of the functions for which the system was designed, and each acting in accord with other systems in the physical world.

The primary differences between earlier systems and the modern systems that we encounter today are *complexity*, *size*, *degree of mechanization*, *speed of action*, and particularly the degree to which *automatic controls* are necessary to their successful operation.

In early mills, where water was the primary source of power, it was not difficult to construct and operate the system. There was a building and a mechanism consisting of water wheel, gears, grinding stone, and bagging room that were designed to act in concert to accomplish the task of grinding and packaging grain. The system was straightforward and simple in design. The water power produced a relatively small amount of usable energy and was fairly constant in its delivery, being governed by the relationship between the movement of the stream and the size and design of the water wheel itself. The gear mechanism worked in one direction and turned one wheel over one surface as grain was poured into a cavity that ground it to flour as it made its way through the system. Bagging was a matter of waiting for each sack to be filled, then replacing it with an empty bag while the full sack was sewn shut, stacked, or stored prior to delivery. If anything were to go awry, the mill was shut down, repairs made, problems corrected, or components changed.

Compare this with a textile mill after the advent of the steam engine. There is now considerably more power available, and it can be delivered at a much higher speed, utilizing more complicated delivery systems consisting of belts, shafts, gearing systems, and other means of mechanical translation. Control becomes more difficult, and reaction speed is greatly reduced. Automatic governors become necessary along with valving and pressure gauges to ensure a safe and constant delivery of power. Safety hazards multiply in the more complicated environment where many machines are connected to the power plant through pulleys and drive shafts. If an error occurs, it must be corrected immediately, and errors are more serious by virtue of the reduced reaction time to changes in the system.

Compare both of the examples above with the operation of a modern supersonic fighter. Reaction time is now so short than an individual pilot is unable to keep up with the changes that are needed to correct for variations in performance. The machine flies itself. It doesn't have time to wait while a human makes a decision. Failsafes and servos do the job once done by human command. The power being supplied and the rate at which it is supplied are beyond the capacity of human pilots to control. The machine is a complex system of complex machines consisting of complex components produced through the expertise of specialists from many fields of endeavor.

This progressive complexity and speed are present in social systems as well. Compare a feudal system with a modern socialistic–capitalistic society. Under feudalism, there were two classes of citizens: the lords and the church, who were the landowners, and the peasants, who were landless and dependent on feudal masters for sustenance and protection. The main concern of the lords was the maintenance of property by defense and by accumulation of wealth. The main concern of the peasantry was survival in a world where their very existence depended on the protection and good will of their sovereign lord. It was a simple system. There were few changes taking place (the major one being the appearance and spreading use of a more efficient plow), and there was little impetus to change. The lords maintained the status quo for their own benefit, and the peasantry maintained it from lack of choice. It was not until the rise of the middle class, those who were landless yet who possessed skills that could be traded for wealth (making them tradespeople) that we see any changes occurring in the feudal system. When this change occurred, it brought on the Renaissance.

A modern Western society is far more complex. The characteristics and distances among classes are quite indistinct by comparison. The variety of activities available to the people, the variety of social structures through which people interact, the variety of goods available, the richness of the culture, communications, transportation, political points of view, and educational opportunities all far surpass those of the feudal system. With this complexity comes the price of a fuller, more extensively intertwined network of social factors. Rather than a single family structure, there are many types of structures. Rather than a single choice of economic endeavor, the availability is almost endless. Rather than a limited choice of where to live, how to live, and what to buy, the variety is staggering. In such an environment (system), control becomes a problem, requiring more extensive networks of courts, police systems, and in general a higher degree of structure in human affairs. This is part of the opportunity cost connected with choosing a modern way of life.

Understanding such a system seems like a monumental task, but fortunately it can be eased by using a systems approach to the analysis of what is taking place.

To understand a system, it is necessary to understand the subsystems of which it is made. To understand these components (subsystems) involves

1. Enumerating and understanding the properties of the component in order to determine the cause-and-effect relationships that make up the next larger system of which the component is a member.
2. Limiting the number and level of components investigated to no more than are necessary to understand the workings of the system of which they are members. Reduction of the system to interacting subsystems should take place only as far as is necessary to give the information about how the components operate that is cogent to our study of the larger system.

This reductionism, as it is called, should only be carried as far as is absolutely necessary. In essence, the investigator is attempting to reduce a given system to a reasonable number of subsystems or components in order to facilitate analysis. Too much reduction results in an unwieldy and useless framework. We must be cautious not to reduce the system so far that we lose sight of our original purpose, or to the point that analysis would become so involved that we experience not seeing "the forest for the trees."

If, for example, I wish to investigate the system known as a personal computer, my investigation may lead me in different directions, even within the determination of subsystems, depending on the *purpose* of the analysis. As a programmer, I might choose to consider the computer as made up of a central processor, the primary storage, a logic unit, input–output devices, and so forth. From here I might move to the ways each of these elements figure in the scheme of developing programs, that is, how does each affect the input of information? What languages are available? What commands place information in certain parts of the main storage? What output devices are available, and how will that affect the way I write the program? Is this an interactive system where I can communicate with the screen as the program runs, or is it passive, where the program is run in toto, followed by output without the benefit of communication during the run?

If I were approaching the personal computer from the point of view of an engineer, the analysis might be quite different. In this case, the important issues may be ones of hardware, exclusive of software. What components are needed to create main memory? How many chips? Which ones? What is the mode of interface between computer and peripheral? Are data ports parallel or sequential? What types of ports are involved? What is the computational rate? What type of power supply is required?

Note that the way we subdivide the system depends on the purpose we have in mind. It depends on what we wish to learn about the system's operation. Also note that we need take this "reductionism" only as far as necessary to perform the analysis in which we are interested. I do not need to know the design of a chip to know that it will store binary information. I need not know the chemical makeup of plastic to know that a plastic key functions as an input device to input a given symbol.

The main considerations in describing the components or subsystems of a system are *what is our purpose*, that is, what is it we wish to learn, and *what level of component reduction is needed to describe the way in which the system functions?*

The Analysis

The system is a set of components that interact in a given set of ways called *criteria*. To analyze a system requires an understanding of the nature and structure of the components and the way that they interact with one another. In fact, the purpose of systems analysis is to observe the functioning of a system and be able to predict the properties that the system has. It must be remembered that the properties of a system are not something innate in the components, but rather a function of their interaction. That is, it is the manner in which the components interact that gives the system its properties, not the components themselves. People are components in some systems. As people, they have capabilities and limitations that make them systems themselves, but the way they contribute to the functioning of a system (for instance, a business organization) is the result of how those human capabilities are used *in concert with the other components of the system* to achieve the purpose of the system of which they are a part. What the analyst seeks to do is to first find the properties of the system that are predictable, repetitive, and therefore considered likely to remain the same within the system's environment. In other words, to find the cause-and-effect relationships among the system components. Second, the analyst seeks to determine in what ways the system is dependent on its environment, find interenvironmental relationships, and predict possible changes in the system as a result of future changes in the system. Once this is done, the systems analyst will further attempt to predict the performance of the system under conditions not statistically determinable, which is, at best, an act of visualizing future trends and is, at worst, an act of faith.

Analysis and purpose.   To properly analyze a system, it is necessary to understand the purpose of the system and then to understand the ways in which the system goes about the achievement of its purpose. Basically, all systems have as their central purpose the performance of some function or functions in such a way that they remain in a state of dynamic balance, either through maintenance of the status quo (steady-state systems) or through growth (dynamic-state systems). Starting with this premise, analysis is merely a matter of determining what the end results of the system are *supposed* to be, the methods used by the system to achieve those end results (interaction of components guided by defined limiting parameters), and the nature of the problems that can arise in the process.

A universal model of a system is presented in Figure 6-1. It represents the basic necessary elements of a system in order for it to achieve and maintain balance. Without each of these components, the system not only does not function, it does not technically exist as a system at all!

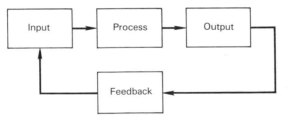

Basic Systems Elements

Figure 6-1 General Systems Model

The *input* is necessary to supply the system with whatever is needed to perform its appointed tasks and achieve its desired goals. Inputs can range from information and pressure from the exterior environment, as in a computer network, to physical resources, such as raw materials, labor, energy, and capital, as in a textile plant. Whatever the form, if there is no input, there will be no activity, since this is the trigger or impetus that starts and maintains the system.

Process is a matter of the transformations that take place within the system by which the inputs are changed, altered, used, or otherwise utilized to achieve the system's purpose. This could range from the utilization of data for creating financial reports or statistical conclusions to the utilization of tungsten, copper, phosphorus, glass, and other inputs to create light bulbs. In whatever manner, the inputs are utilized, and by some process, the system functions, its components interacting to create some desired result.

Output is the purpose of the system. The output is the thing, idea, object, product, or condition that the system is designed to achieve. This represents the *raison d'être*, the reason for existence, of the system. Without the output, there would be no purpose for its being, and the system would simply cease to exist.

Feedback is the control mechanism by which the system determines if it has achieved its purpose. Changes in the environment within which the system is operating will be subject to change over time. In any real-world situation, this is so highly probable that it is often considered a given. The existence of the possibility of change means the necessity within a successful system of a method of correcting activities to adjust for that system. The primary purpose of the feedback element is to supply the system with information so that it can correct itself. Feedback may be either passive or active. It may be part of an automatic system designed to always create certain changes in input and processing behavior as a result of certain criteria, or it may be discretionary, in that it supplies some other decision-making element of the system with raw data upon which a decision may be made. The governor on a steam engine automatically puts on the brakes if the speed of shaft revolution passes a certain critical point. In a market system, the feedback on sales made in the market is just a single element in decisions of production, distribution, and advertising. Both are cases of feedback.

Given this basic model of a system, the analyst needs to determine what the system actually does (how it functions). One very efficient means of doing this is through determination of input-and-output relationships to the process. The analyst seeks to answer questions concerning these elements, usually beginning with the output and working backward. For example,

1. What type of output can be expected from the system?
2. How does this output reflect the purpose of the system?
3. How fast is this output generated, and what is the feedback time between output and effect of interaction with the surrounding environmental system?
4. In order to create the output, what forms of input are required?
5. How fast can the input be received by the system?
6. How fast can the system change its rate of input reception in reaction to feedback information?

7. How does the system process the input to create the output?
8. In what ways can the process fail?
9. What are the probabilities of failure of any systems component at any time under given circumstances?
10. What are the probable consequences of each type of failure (a) to the system? (b) to the environment?

The answers to these questions are available either through analysis of historical data, through experimentation, or through inspection. The last element, inspection, is really a special case of experimentation. If experimentation is the necessary form of investigation, measurement and analysis using the scientific method should be undertaken.

HISTORICAL APPLICATIONS OF THE SYSTEMS APPROACH

The concept of systems is not a new idea. It has been with us throughout history, from the Greek division of matter into fire, water, air, and earth (all matter being considered to be the result of various interactions among these four basic elements), through Leibniz, Hegel, Hesse, and Kohler. The concept is evident in the Gestalt school of psychology first developed in Germany, which describes a person's psychological makeup as the result of all things experienced by the person acting in concert to create a given set of response patterns to external stimuli. It is also apparent in the scalar approach of Henri Fayol, who first enunciated the principles of authority and responsibility and who developed the concept of delegation of authority through a well-structured, hierarchical organizational system. It is apparent as well in the concept of bureaucracy, in the taxonomic classification of life on this planet, and in the cybernetics of the twentieth century and the synergetics of Buckminster Fuller. The concept of a system has long been apparent in humanity's attempt to understand itself and its world, yet it was not until the twentieth century that it was dealt with as a system.

Consider the hierarchy of large corporations. The principle of the hierarchy still overshadows other organizational patterns with its highly structured format. Each job is listed in its proper place (component), designed to carry out certain specific functions (making each component a subsystem in itself), having interaction with all of the other positions in the firm according to highly defined relationships (defining the procedures through which the components perform the process of the system). Inputs in the form of money, market reports, raw materials, and information flow into the organization where they are used to produce a useful product (process to output), which is then sold on the open market for money, which flows back into the system as feedback.

The system operates and controls itself in accordance with its relationship to the environment and changes in that environment, such as changes in demand for the product the system is producing. And each element of the system is defined prior to operation down to the last detail. Hierarchical structures such as this are *planned* and *organized* as the company is put together, then altered in response to environmental influences in its search for what works (seeking balance), and in its efforts to grow as a company (implying dynamic growth).

In a similar manner, the military is structured as a system, with each component down to the individual foot soldier defined as to purpose and relationship to every other member of the system. And the more complex the organization, the more need there is for systems organization, as is evidenced by the degree to which military systems codify and specify activities, positions, and so forth.

TECHNOLOGY AND THE SYSTEMIC APPROACH

The systemic approach is a logical one to take in an investigation of technology and society. Technology is a subsystem of what it is to be part of humanity. As already seen, the artifacts that we create are manipulations of physical laws to allow us to structure our society the way we want it to be. In the same way, the social elements of our society are an integral part of what it is to be human, defining the ways that we choose to interact, cooperate, structure groups, and function in dynamic social patterns. This is another subsystem of humanity.

By virtue of this connection, the relationships between social structures and technology can be studied in an orderly fashion using various statistical and methodological techniques within an overall framework of systems analysis. The meat of this pie, however, must wait until we have explored a few techniques of analysis in subsequent chapters.

CONCLUSION

We live in a world controlled by natural laws. These laws describe the ways in which individual elements of our physical environment interact with one another. Understanding these relationships is a matter of determining cause-and-effect patterns. Through understanding these patterns of behavior we are able to predict what will happen in the world with some degree of accuracy. This is the way we make assumptions, and on these assumptions, we base our behavior.

The cause-and-effect relationships that we encounter are descriptive of a general overall pattern that involves the entire network of physical existence. Each physical element in the real world is connected to every other element in that they are all part of a single system. We can define a system as a collection of components connected to one another by various interactive procedures. Each system consists of smaller subsystems that, in turn, consist of still smaller subsystems down to the smallest component imaginable, to the world of subatomic particles.

The process of analyzing phenomena by breaking down systems into constituent components is known as the *systems approach*. Breaking down systems into subsystems and following this with the same procedure for each subsystem is known as *reductionism*. It is a method of reducing problems, perceived subsystems, or organizations to a size that makes them manageable and predictable. The purpose of systems analysis is to carry on this reductionism to the point that a system may be analyzed and understood in conjunction with one's purposes. Reductionism should only be carried out insofar as it accomplishes one's purpose, further subdivision being meaningless.

The methods by which one carries out systems analysis are useful in the study of the relationship between technology and social structures, as both technology and social structure are subsystems of the larger system we call humanity, which is, in turn, part of the larger yet world system, and are therefore subject to reductionism as part of those larger systems. Armed with the methodology of statistical analysis and predictive modeling, it is possible to use this approach.

APPENDIX: EXAMPLE OF SYSTEMS APPROACH IN USE

As an example of how the systems approach may be used to analyze a system and predict outcomes on the basis of that analysis, the following simplified view of a macroeconomic system is offered. For the sake of brevity and to avoid lengthy explanations, the system has been kept to a simplified form.

The purpose of the analysis is to investigate the nature of a macroeconomic system, that is, a national economy, in hopes of developing predictive data on the probable performance of that economy under various combinations of conditions.

Definition of the System and Its Purpose

An economic system is designed to produce and distribute goods and services produced from scarce resources, so that those goods and services may be consumed by the final user. The *purpose* of the system is seen as the creation of satisfaction among consumers as a result of consuming goods and services.

Outputs

Goods and services can be defined as final goods and services in the form of produced items of value and valuable services to the public that consumers would wish to consume.

Inputs

Productive resources can be defined as natural resources (land with its original fertility and mineral deposits), labor (the productive efforts of people working with their minds and their muscles to create useful products), and capital (all manufactured productive resources, including machinery and equipment, buildings, and improvements to land).

Process Components

The economy is considered to consist of three sectors and three markets, the former consisting of industry (or the investment sector), consumers, and government; and the latter consisting of the product market, the factor market, and the credit market.

The investment (or industrial) sector.   This sector consists of industry that produces goods and services for the general population through the product market in hopes of receiving profits for its efforts and therefore increasing their overall economic welfare. The industrial sector also purchases the means of production through the factor market from consumers who own said means as private property, so that they may have the means by which they create the goods and services to be sold.

The consumer sector.   This sector consists of individual consumers in the economy who purchase goods and services from industry through the product market for the purpose of receiving satisfaction through the consumption process and therefore increasing their overall economic welfare. In addition, consumers offer their property and labor to the industrial sector through the factor market, so that they may receive compensation in the form of rent, interest, or profits, and thus have the means by which they can purchase finished goods and services for consumption.

The government sector.   This sector consists of all governmental entities, collectively known as the *public sector*. It receives taxes and borrows money from both the industrial sector and the consumer sector and utilizes these funds to provide goods and services to the population as a whole, based on the desires and needs of the members of the society. It purchases goods and services through the product market from the industrial sector and purchases labor and other factors of production from the consumer sector through the factor market, in turn supplying public goods as needed.

The product market.   This market consists of all exchange mechanisms through which the goods and services of producers are bought and sold.

The factor market.   This market consists of all markets where the means of production (natural resources, labor, and capital) are bought and sold.

The credit market.   This market consists of all institutions that function as intermediaries between consumers who save and industry and government that borrow.

Determination of Component Functions and Interactions

As indicated in Figure 6-2, each of the three sectors interacts with the other two sectors through all three of the available markets. This diagrammatic description of the processes inherent in the operation of the macroeconomic system gives an overview of the types and forms the interactions take. A detailed analysis of each of the interactions yields a more complete model of the system.

Consumer interactions.   The determination of consumer behavior is predicated upon two underlying conditions—the availability of income for expenditure (so-called disposable income) and the manner in which this available income is spent. To determine the amount of disposable income, the tax rate must be known, yielding a formula, $D = (Y - t)$, where D is disposable income, Y is total income, and t represents net taxes, here assumed at an average flat rate to avoid unnecessarily complicating the issue. In addition, it is necessary to know the value of the marginal propensity to consume (MPC), which is a measure of the tendency of the average person in the economy to spend a portion of the next dollar earned. A final necessary value is the amount of autonomous consumption in the society, that is, the minimum level of consumption that must take place in order for the population to survive. This factor is a discrete number based on historical data concerning prices and the level of consumption necessary for survival. It is usually represented as the constant a. Combining these factors, we find that the *consumption function* is described by the formula, $C = a + b * (Y - t)$.

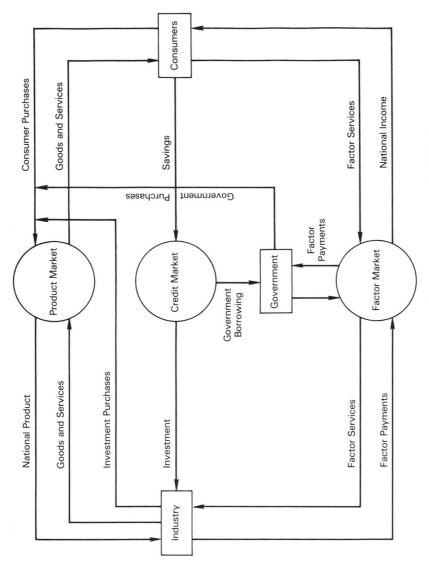

Figure 6-2 The Circular Flow of Goods and Services in the U.S. Economy

Investment function.   The investment function is the result of two relationships, one concerning the level of expected return on investment (ROI) and the other involving the interest rate. The level of investment within an economy, that is, the amount of money that producers are willing to invest in the production and selling of goods and services, is inversely related to the interest rate (the higher the interest rate, the more costly it is to borrow funds to invest) and directly related to the expected rate of return on investment (since the greater the ROI, the greater the expected profits from the production and sale of the products in question). This function can be denoted by the formula, $I = c - dr$, where I indicates the rate of investment, d is a constant of proportionality, c is the ideal level of investment, and r is the prevailing interest rate in the economy.

Government spending.   Government spending is defined through the political structure, being indeterminate in the normal sense. The estimate for government spending utilized in economic modeling is based on the official or expected federal, state, and local budgets. In formula, it is expressed as G, where G is government spending as specified by the budgets.

Combining these three sector actions to define the aggregate demand in the economy, we find that $AD = C + I + G$, or $AD = a + c + b * (Y - t) + G + c - dr$.

A secondary relationship among the sectors of the economy involves their relationship to the money supply, which determines prices and interest rates. The guiding rule here is the fact that the quantity of money demanded in an economy is inversely related to the nominal interest rate and directly proportional to the level of nominal national income. This is called the *theory of proportionality* and is mathematically represented by the formula, $Md = e * Y - f * r$, where Y represents nominal national income, r represents the prevailing interest rate in the economy, and e and f are constants of proportionality.

One final relationship must be noted—the determination of the equilibrium level of nominal national income, that is, the level of income at which the economy is balanced (in equilibrium). It is denoted by the location of the intersection of the aggregate demand curve, defined above as AD, and the aggregate supply curve, given by the formula $E = Y$. Combining the two equations and solving for Y (in this case, a value found on the x-axis), the equilibrium level of nominal national income is located.

Armed with these relationships, we can therefore define the operation of a simple macroeconomic system as the solution set of the two equations, $Md = e * Y + f * r$ and $AD = a + c + b * (Y - t) + dr + G$, such that the values of r and Y are identical for both formulas. When Y and r are the same for both formulas, the result is a macroeconomic system in equilibrium (that is, it is balanced).

This brief synopsis of a simplistic solution to the macroeconomic model should serve to indicate the complexity to which systems are capable of moving in a very short time. It is quite beyond the scope of this book to explain in detail how to construct highly sophisticated systems models. However, as indicated in the next chapter, modeling is a primary source of analysis for quantifiable systems. A more reasonable and less mathematical approach to the analysis of technological social systems is presented in a later chapter.

THOUGHT AND PROCESS

1. Anything that is a process can be thought of in terms of a system. It is the fact that the systems approach is so universally applicable that gives it its strength. Choose some activity or process in which you are regularly involved, and use the systems approach to analyze it. Some examples might be (a) the process by which a person makes decisions that get him or her up in the morning and to work on time, (b) the process of washing the car, (c) the system known as "Thanksgiving dinner," (d) the process through which one determines what to do on a Saturday night, or (e) the system known as the family.

2. Many systems have specific characteristics that are instrumental in making them what they are. For instance, the only characteristic that maintains a business organization as what it is lies in its purpose, that is, the objectives that it develops to accomplish. Without these objectives, there is no system, no matter how much the physical parts of the former system still exist. With this in mind, analyze the following systems in terms of their distinguishing elements, classifying them by their input, process, output, and feedback elements:
   (a) the human body
   (b) an oak tree
   (c) an automobile manufacturer
   (d) the solar system
   (e) the play, *Hamlet*

3. If every action taken is the cause of some other occurrence, which, in turn, becomes the cause of other effects, what does this imply for the independence of events in the real world? That is, if we carry this to its logical conclusion, are we saying that all things are really interconnected to one another? Why? Why not?

CHAPTER 7

Modeling, Simulations, and Gaming

Charles McAfferty was an unlikely candidate for the kind of success he had experienced. He was a quiet young man, slight, with straight dark hair and deep blue eyes. He was, all in all, quite unobtrusive and passive to the casual observer. All of his life, he had been the favorite of his friends, entertaining, witty in a droll sort of way, and he had gained the reputation for being the least offensive yet kindest of people. But when it came to business, Charles was all business, and that's a fact.

Anyone who knew him well was aware of what to expect when Charles was confronted with the opportunity to make a profit. His entire personality changed. His priorities suddenly shifted. No one was safe if he got in his way in a business deal. He had been in jail six times because of his willingness to take extreme risks for the opportunity to add to his considerable holdings, and he had not always been successful in extricating himself from legal difficulties without a considerable expenditure of time, money, and effort. But to him, it was all just a part of the game. Nothing was more important than his financial empire.

Starting with only a small nest egg, Charles had been able to amass huge holdings and incredible wealth. His shrewd purchases in real estate, his uncanny ability to shift in and out of certain stocks at just the right moment, and his fearless aggressiveness in the face of stiff competition had led him on a road to success that few have ever been able to equal. Charles was truly an amazing person.

He sat at the conference table, his small thin hands folded in his lap, a faint shadow of a smile on his face, waiting for some response from the man sitting opposite him. His eyes gleamed as he thought of the irony of it all. Charles thought of all those people whom he had relieved of their fortunes. All those others, so confident and so sure of themselves and who were now completely wiped out by his chicanery and his greed, meant nothing to him, and never had. But the man across the table from him was something else again. This man would be the supreme victory.

Charles was on the verge of his greatest triumph, his most incredible deal, and it was to be at the expense, literally, of his best friend.

Just think of it! Tom Hendrix' fortune was within his grasp. He had been a rival since they were small children, the older brother he had never had, the hero of dreams as long as he could remember. And now he was sitting directly across the conference table, only moments from financial ruin, and it was all Charles McAfferty's doing! Victory would be sweet indeed this day.

Tom, on the other hand, was anything but pleased. He sat silently brooding, his crossed brow reflecting his dismay at his impending financial doom, his folded hands resting lightly on his thin lips, as if to say, "Shhh! Don't bother me while I'm thinking." He shook his head and recalculated his position one more time to be certain. He knew in his heart that he was ruined, but he simply could not admit that Charles had finally destroyed him. It was just too horrible to believe. He had to look one last time for an out, some loophole that would save him from his opponent's avarice. But it was useless. Charles had him, and he knew it.

"Ready to admit that I've got you, Tom?" he asked a bit too casually.

Tom looked up at him. His gaze would have been more than a weaker person could endure, but Charles merely broadened his grin a bit and waited.

"You really think you've got me this time, don't you Charley?"

"I know I have, buddy. Why don't you just admit it and give in. I'm about to destroy you. I'm about to own everything you've spent so long accumulating. I'm going to take the shirt off your back. More than anyone else in this town, I've wanted to beat you, Tom. And now I'm going to have it all."

"Probably," Tom said noncommittally. "Tell me something, Charley. What's so special about me? Why go after me? I've seen you run people out of business before. I've watched you take them for all they're worth and pull them to their knees, but it was never personal with you. For you it was always just mechanical, part of the fun, a way to pass the time, to garner victories, and to add to your fortune. But with me it's different, isn't it? Oh, you needn't answer, Charley. I know it is. This is personal with us, and we both know it. It's not the money this time, or the holdings, or the power. You've already got more of that than you can possibly use. This time it's just to beat me and for no other reason. Why? What's so special about me?"

Charles released a sinister and somewhat uncharacteristic laugh. He leaned forward and locked Tom's eyes, the steel blue of his own turning even colder as he spoke.

"I could lie to you and just say that winning is the name of the game, but I won't. You're right, of course. It is personal. I've waited a long time for this, Tom. We're terribly different, you and I. You're all caught up in your own self-righteousness. You remind me of a character in a romantic novel, naive and full of goodness, and all that rot. I've watched your empire grow just as mine has. I've watched you lend money to people when they needed it, to forego payments when borrowers couldn't afford to pay. I've seen you give away your wealth to subsidize the folly of those who dared to compete against you. And I've watched them bail you out of a few tight places, too. It was pitiful to watch, Tom, and pitiful and weak of you to allow it.

"Do you remember the last time I was in jail? Did any one of them offer to give me a hand? Did any of them offer to put up my bail or pay my fines until I could get back on my feet? Oh, you did, Tom, but you must have known that I could never accept any help from you. Not my childhood hero. I did it all on my own. I got myself out of my own scrapes, and I went on to watch every one of them fall by the wayside, ruined one by one, wiped out by my hand. They're all gone now, Tom, all but you. And now I've got you. I'm going to really enjoy watching you going under, pal. It's going to be a pleasure to wipe you out the way I did the others. I've won, Tom, and no one can take that from me. Why not just admit it? I've got you and you know it."

Tom shook his head. He took a deep breath and released a long, soft sigh. "If you want my holdings, my dear friend, you'll just have to take them. I'll not give in to you that easily."

"So be it!" Charles hissed. "Let's get on with it then."

Tom sighed again and reached out his hand to the conference table. He picked up the dice and rolled. Any roll but a seven was going to mean disaster. Only a seven could save him from financial ruin. And a seven it was! Tom counted out the seven spaces on the board and landed on Marvin Gardens, the only space on that side of the board that was his.

Charles grabbed the dice and rolled with a vengeance, then advanced his token ten spaces. He groaned when he realized that he had landed on Park Avenue, another of Tom's properties. Piled on one side was a neatly arranged stack of five hotels! He had done it! Tom had wiped him out in a single throw of the dice. His financial empire was toppled, his wealth drained off in a hideous orgy of rental payments! Tom had won the game!

Tom patted Charley's shoulder as the slight young man slumped, head in folded arms, among the scattered playing pieces on the board.

"Don't take it so hard, buddy," Tom said. "After all, you're the one who wanted to change the rules. It could have been just one hotel over there. And it had to end soon anyway. I've got a class at 8:00 in the morning and it's already 2:30 in the morning now. We'll try again tomorrow afternoon at my place. Cheer up, pal. After all, it's only a game."

Gotcha! Or perhaps not. By this time, it should be evident that things are not always exactly what they appear to be. And in a sense, it is that very proposition with which this chapter concerns itself. How does one determine when something is behaving as expected? How does one define what the world looks like? How does one gain experience in formulating an understanding of the real world? How can one learn to predict what may happen and how it will happen under a certain set of circumstances? All these questions can be dealt with in terms of models and their adjuncts, simulation and gaming.

A small child picks up a toy gun and begins shooting imaginary soldiers. Another child holds its favorite stuffed animal and tries to feed it a piece of apple, frowning in displeasure because Jojo isn't hungry. Scientists peer through small windows or watch closed-circuit television broadcasts to study the passage of smoke streaming across an experimental airfoil in the face of a windtunnel hurricane. A business executive punches yet another set of numbers into a desk computer and watches as the profit estimates for next fall change drastically. These are examples of different types of models being used to study and to mimic the real world. Their use is a phenomenon that we all experience every day and in many ways.

How can things that are so different all be examples of the same technique? What is the common thread among stuffed animals, toy guns, mathematical formulas, and airfoils? Each is an idea of the real world, designed to be tested for its accuracy. The more accurate a model may be, the more useful it is in *describing* and *predicting* what takes place around us. The diversity of models is no more extreme than the diversity of the real-world phenomena one might wish to study. In each case, the model noted is *appropriate* to the type of phenomenon under study and the type of results one may be interested in achieving.

A *model*, then, is a copy or imitation of a physical structure (a thing) or a concept that is designed to demonstrate certain characteristics of that thing or concept in accordance with the purposes of the modeler. In this context, the form the model takes is a matter of how one wishes to use it. An aeronautical engineer would not use a plastic airplane model to study the effects of friction on the surface temperature of a metal alloy airplane wing, as this would be an inappropriate medium for such a test. Nor would he or she use a series of formulas describing aerodynamic behavior as a means of teaching pilots the "feel" of an aircraft in a steep dive. In the first case, the plastic model is designed to *look* like the real-world object, not to *behave* like it; in the second, the formulas are symbolic representations designed to relay information about performance characteristics (descriptions), not to artificially create physical experiences for pilots. The choice of models is therefore important if we are to find them useful. In this sense, a major factor in picking what models to use is *appropriateness.*

FORMS OF MODELS

The form that a model takes varies depending on need, just as the choice of models varies for appropriateness. Generally, models are said to be an *analog*, an *icon*, *verbal*, or *mathematical*. This classification embodies the basic types of models that are encountered in everyday life.

Analog Models

An *analog* model is a model that behaves in a way similar to the reality which it is designed to represent. A model airplane that actually flies is an example of an analog model. The model behaves similarly to an actual airplane in flight, being subject to and reacting to the natural forces that a full-size airplane would be expected to encounter. Performance characteristics are approximated using this type of model, and the analog model offers an inexpensive and, in this case, safer alternative to producing a prototype at the outset.

Another example of an analog model is the slide rule, a device that translates mathematical relationships into spatial patterns. The movement of the slide along its track and the distance covered are directly analogous to the change in values experienced when one uses logarithmic principles. Adding logarithms is represented on the slide rule by adding lengths, and the results exactly mirror the behavior of the numbers in question. This is the same procedure that is done when clocks are used to measure time, a nonspatial concept, by sweeping out lengths of arc on the clock face, or the use of an orrery, a mechanical model of the solar system in which balls connected to lengths of rod are geared to circle a central hub indicative of the sun to show how the planets sweep out arcs through time. The movement of the balls around the hub is analogous to the movement of the planets around the sun.

Analog models are extremely useful in investigating and understanding physical phenomena. They often produce large amounts of information in their creation, as the modeler strives to mimic the real world. By going through the thought processes and the activities necessary to produce the model itself, the modeler learns about the mechanisms and behavior of the thing or concept that he or she is attempting to describe. Many airfoil designs, for instance, are tried and rejected before a final successful design is achieved for a new aircraft. It is through the ability of the analog model wings and fuselages to be studied and modified that we are able to learn what will work by correcting what does not work in the model.

The use of analog models in industry should be quickly obvious. New devices are first produced in smaller, less expensive form to allow for refinement of original designs until the producer has a useful, workable "model" that can be manufactured and made available to the public. Yet there is a less obvious place for analog models in the field of technology, particularly as technology applies to the environment and other elements of the social structure. Analog models can predict the effects of erosion on soil. They can be used to telescope time in the study of populations and population control (as we have seen in the case of Malthus). They can be quite useful in studying the effects of architecture on earthquake safety or wind and weather in the inner city. To relegate their use solely to the production of technological innovations would be shortsighted in the extreme.

Iconic Models

Iconic models are somewhat different in that they are designed to look like (resemble) the physical reality that they are describing, but not necessarily to behave in a similar manner. Toy trucks and animals are examples of icons, as are architectural models of houses and office buildings, or the previously cited example of the plastic airplane model. None of these models necessarily behaves in ways indicative of what they are describing, but they do resemble them physically. Paintings, sculptures, and many computer-created illustrations are other examples of icons which are designed *only* to look like the "real thing."

How are they useful to technology? Icons can be utilized to study aspects of real-world phenomena that are not directly related to performance, though every bit as important. Industrial design uses icons to discover the most aesthetically pleasing form for useful products to take. The shape, color, or texture of a product may have little or no effect on how well it performs the task for which it is created, but it may make the difference between success or failure in the marketplace as much as shoddy workmanship or poorly chosen materials can.

Spatial relationships can also be studied using icons, as in the homemaker who uses models of furniture to rearrange a living room many times before deciding on the best combination for comfort and function. Large companies often use iconic representations of machinery and equipment in conjunction with physical models of new plants in order to best determine how these items should be positioned to maximize efficiency and minimize hazards.

Yet these examples are only a small part of the huge number of uses to which we put icons. The majority of them, and those that have the greatest effect on our society, are in a special category known as *toys.* Through the use of toys, children and, to a larger extent than you might imagine, adults practice the skills that they will need later for survival in our society. Iconic toys in combination with a fertile imagination allow children to develop and walk through numerous scenarios of possible and expected futures without risk to life and limb or fear of suffering trauma. With this in mind, it should be realized that the actions and interests of adults are at a minimum partially determined by the icons that they dealt with as children. The child who plays computer games today is already intimately familiar with and comfortable with computers as an adult. The child who plays with toy trucks and cars is more aware of the hazards of these modes of transportation, their capabilities, and their possible uses. Children practice through play, increasing their ability to deal with the adult world once they find themselves faced with the responsibilities and opportunities of that world.

Verbal Models

Verbal models are descriptive in nature. They are designed to convert thoughts and concepts into language to establish relationships and restrictions of the real-world system and to organize them. It is in this establishment of relationships and restrictions and in their organization that verbal models excel.

In order to create a verbal model, it is necessary to *conceptualize* what is being modeled. A process must be undergone through which the modeler structures his or her understanding of the phenomenon being described and organizes the information about the phenomenon into some logical pattern. Verbalization forces inspection. Describing something, as for instance in a simple definition, means that the person must decide, first, what is included in the definition, second, what must be specifically excluded from the definition, and third, what sort of logical structure to put the definition into. By way of example, notice how this process is illustrated by entries in any dictionary, then read the instructions provided with your favorite game or the instructions for assembling the lawnmower you recently purchased and are still trying to figure out, or, for that matter, take a second look at the structure of this book. It is nothing more than a verbal representation of an idea about how the world is put together, how it functions, and how to study that functioning.

Verbal models have one inherent weakness, and that is interpretation. They have a tendency to become ungainly due to the necessity of exactitude of description to produce accuracy of description. The old saying about a picture being worth a thousand words has been around for a very long time, and its survival stems from it being true. Words have a tremendous power to describe and to describe accurately to the smallest detail, but if the interpretation of the words used varies, the meaning suffers. Simple concepts take few words to describe. Complicated concepts require paragraphs, chapters, indeed volumes to adequately explain. When are they useful? When we need to include, exclude, and organize our ideas about real-world phenomena, and to offer knowledge in a way that can be easily understood by a large number of people. Verbal models greatly enhance the dissemination of knowledge and, along with it, understanding.

Mathematical Models

The fourth form of models is the mathematical form. *Mathematical models* are manipulative, symbolic representations of reality designed to describe the *relationships* among certain factors of a thing or concept, to establish restrictions (limits) on the thing or concept, and to use behavioral characteristics to *predict* with some degree of accuracy the manner in which behavior will change under given sets of circumstances.

Mathematical models are usually in the form of a formula or set of formulas describing and predicting behavior. Much of the information we have about real-world behavior is in the form of formulas. The physical sciences, such as astronomy, physics, and chemistry rely heavily on this predictive and descriptive form of modeling to interpret what is observed. In these sciences particularly, the behavior that can be quantized (that is, reduced to numbers, ratios, and so forth) is the behavior that can be investigated, studied, and used. Technology is heavily dependent on mathematical models, as are the social sciences, which utilize the concepts of statistics to take nondeterministic phenomena and generalize them into manipulative form. Mathematical models are among the most successful at predicting macrocultural and microcultural behavior. Examples are found in nearly every discipline, from the models of market behavior encountered in economics to the laws of electrical behavior in physics.

THE MODELING PROCESS

Fortunately, there is a simple and general process by which modeling can be done. It is equally applicable to all types of models and involves determining what type of model to utilize and how to go about it. The process is presented below in brief.

The Modeling Process

1. Gather information about the concept or physical structure to be modeled.
2. Based on this information, reach conclusions about the nature, characteristics, and behavior of the reality to be modeled.
3. Determine an appropriate form for the model; the degree of detail required; what elements are important in understanding the nature of the reality; and, of those elements, which should be included in the model itself.
4. Build the model.
5. Compare the model with reality to determine the degree to which the model actually approximates the reality.
6. Adjust the model as necessary to achieve the desired "fit."

It may be noted that there is a distinct similarity between this modeling process and what is known as the scientific process used in the physical sciences. This is not too surprising, since most scientific theories and laws are models of reality. All that has been done here is to present a more generalized format for the application of the technique to a less strictured range of problems. For clarity, the steps are discussed in somewhat more detail below.

1. *Gather information about the concept or physical structure to be modeled.* A model can be rendered useless if this step is not done properly. The ignorance of information can change the entire complexion of a problem and greatly influence the direction that an investigator takes in creating his or her final concept of reality. It is imperative that all possible information pertaining to the aspects of the reality under scrutiny be gathered for study. A single fact can totally change the meaning of research by either its inclusion or exclusion. How much research is enough? When is it too much? This is a question that the investigator has to answer. With experience, it becomes easier to make this determination. A certain understanding develops over time to tell the investigator when the law of diminishing returns will set in, yielding less information per unit of effort than is acceptable. In general, research suggests research, and the details present themselves in the course of investigation.

2. *Based on this information, reach conclusions about the nature, characteristics, and behavior of the reality to be modeled.* What did you find in your research? What facts presented themselves? What relationships did you observe? Just thinking about the subject will yield a wealth of possibilities. It is in the combining of the information gathered through research and its ordering into logical cause-and-effect relationships that produces the first mental picture of the reality we are attempting to model. The first efforts tend to be macro in structure, dealing with sweeping, overall statements about the way the reality is structured, the form it takes, and the way it behaves under different sets of conditions. Observations and the observations of others often suggest possible pictures of the reality that can be tested, incorporated into our present concept of what is, and rearranged or modified to fit, yielding a much better understanding of what is being dealt with.

As an example, consider the case of an infant exploring the world of the nursery floor. Toys lay round about, bright and inviting, awaiting inspection. The infant focuses on a ball and reaches for it. The toddler notices its shape, its color, the fact that it shines and reflects light, that it is smooth to the touch, that its weight is not too great to pick it up, and that it can be grasped with both hands and held up, whereas one hand will not do the job. These are all observations that could be considered first-hand research. (Hands-on investigation?)

The infant reaches certain conclusions: So far, the object is not dangerous, that is, it does not illicit feelings of discomfort and pain. It is pleasurable in that it feels nice and the bright red color looks nice. It is enjoyable to learn as much as possible about it. At some point, however, the infant tires of it, being attracted by some other toy, and lets the ball drop to the floor. At this point, the child observes it bounce once or twice, then roll toward the corner of the room by the closet.

The child now has a picture of a real-world object, the red ball, and an idea of the nature and behavior of the ball that can be used later when encountering similar objects. The memory model of the red ball's behavior can help the child to react in appropriate (successful) ways with other spherical objects. Needless to say, the infant will find out that other spheres do not necessarily behave the same way as the red plastic ball, as, for instance, in the case of a bowling ball (should the child try picking one up with two hands) or a tennis ball (more bounce, fuzzy surface, smaller size, less weight). But the child has a model to guide him in supposing what to expect, and it is one that can be modified with experience to include the differences encountered with other similar objects.

In the same way, when we build a model and gather information about its behavior and characteristics, we formulate a mental picture of expected results from encountering the real-world phenomenon that the model represents. The difference between what the child does with the ball and what an adult does with, say, the concept of acid rain, is one of content, not one of procedure. The *process* is essentially the same.

3. *Determine an appropriate form for the model; the degree of detail required; what elements are important in understanding the nature of the reality; and, of those elements, which should be included in the model itself.* Through steps one and two, an understanding of the nature of the reality has been obtained. Using this mental picture, it is now possible to consider what should go into the model that we are going to formally build. The elements to consider involve what is important to the purposes of the investigator, what restrictions exist on the possible nature of the model, and how these affect the final form that the model will take.

By way of example, let us return to the orrery mentioned earlier. As indicated, an orrery is a physical representation of the solar system that demonstrates the movements of the planets around the sun on their yearly journey. The positions of the planets are indicated by small spheres on long rods attached to a central hub and geared to change position in the same ratio as the various planets change position through time. The moon sphere is geared to circle the earth sphere every 28 revolutions of the earth sphere in its axis. The earth sphere is geared to circle the central sun sphere once every 365.4 (approximately) revolutions of the earth sphere about its own axis, and so on. Certain characteristics of the physical system known as the solar system have been very faithfully mimicked in the orrery model, while other obvious and important ones have been ignored. The spheres are not placed at a ratio of distance that is physically analogous to the distances the real planets demonstrate. The composition of material of each sphere is far from commensurate with that of the real world. The central sphere of the model does not consist of mostly hydrogen gas with helium and a trace of the other known elements mixed in. The earth model does not have a liquid iron–nickel core (as some "models" predict the real earth does) surrounded by a rocky lithosphere and mantle. There are no orbiting rings of rock and ice about Saturn or Uranus. All of these elements of the physical reality under investigation have been ignored.

Yet the *spatial* relationships of the planets are maintained to as high a degree of accuracy as the modeler can obtain. And this is because of (a) the characteristics that are important to the investigator, (b) the possibilities, given the structure of reality, and (c) that which appears to be the most logical form for the model to take in order to copy those aspects of the real phenomenon of interest. It is not possible to accurately measure the distances, even on a reduced scale, among the planetary bodies. The distances are too huge and the model would be useless. It is equally impossible on a small scale to construct a physical model incorporating all of the true mineral and chemical combinations existing in the real planets. And why would we want to? The purpose of the model is to show the *movements* of the various planets about the sun, and that aspect of the reality is successfully and clearly demonstrated by the orrery.

Had the modeler wished to discuss distances, he or she would have used different forms of models. Mathematical relationships could have been formulated and arranged into tables or charts and graphs (symbolic icons). If the modeler had wanted to demonstrate the chemical and physical phenomena illustrated by the real-world system, he or she could have formulated the laws of gravitation as Newton did, or dropped balls from the leaning tower of Pisa (as, alas, no one of importance has ever done), or demonstrated the activity of the sun quite handily by detonating a fusion bomb! Copying nature in models is a matter of choosing the model appropriate to the purpose of the model. *What* you are attempting to illustrate will dictate the *form* that your model takes.

4. *Build the model.* If the previous steps have been done well, the actual construction of the model should be a simple task indeed. Armed with an understanding of the nature of the reality, what characteristics

and behaviors are important to the investigator, and the form that the model should take to most perfectly demonstrate those important aspects, construction of the model itself is no more than physically doing what has been planned.

In building the model, many of the problems that have been overlooked will present themselves. No matter how detailed an investigation of a subject, the modeler is likely to forget some aspects of the problem, resulting in stumbling blocks, unforeseen changes in approach, and, most importantly, new understanding about what is being represented. The building process is really a combination of building, discovering, reassessing previous steps, and restructuring of the model itself. Piece by piece, the parts of the model fall together and are adjusted to fit in with the other parts until some cohesive creation exists that the modeler believes to be a representation of the phenomenon being investigated.

Consider the iconic modeler who constructs a detailed ship model. He or she may first search out designs, specifications, and photographs of the ship the modeler wishes to model. He or she may even seek out other models of the same subject, noting as well as possible the techniques of fabrication and detail to assess what can be done and how reasonable it is to expect a given type of model in the end. Proceeding from this, the modeler reaches conclusions about the final design, how detailed the model should be, how large it should be, what materials to use for the various elements in the model—Should small bronze cannon be cast, or should I use painted lead or wood? How about sails? Are normal cloth weaves too heavy to create a realistic effect?—and generally how far to go with the process.

This is all very well and good, but as soon as the modeler begins constructing the model, he or she discovers new information not considered previously. What if the authentic woods are too grainy to be easily worked in the smaller scale? What if thread is not available in the proper sizes and turns? What should be done about catlines too small in the model to be woven and tied accurately? There is a constant adjusting of the concept of the model to changes in the way the problem is understood in order to put the various elements together in a useful way.

An equally valid example would be the structuring of an economic model to deal with "simple" market responses to changes in price. On first inspection, it would appear that an observation of market behavior could create a useful, though crude, model, and that is partially true. The laws of supply and demand are well documented and have been borne out over years of observation. Yet the market process is far more complicated than that, and if the modeler is interested in any degree of accuracy, the list of considerations with which he or she must deal grows rapidly. What about advertising? Are we going to incorporate a model for the effects of advertising into our market model? What about quality differentials among manufacturers? Are we to assume that every producer's product is equally valuable and equally well made? If this is a distortion of reality, how important is it to correct that distortion? Will it affect the market model at some later, critical point?

These examples illustrate the value of the model construction to the modeler and to the understanding of any reality. It is in the process of the model construction that the modeler can gain a true sense of the nature of the object of study and is able to satisfy the initial motives in investigating the phenomenon to start with.

5. *Compare the model with reality to determine the degree to which the model actually approximates the reality.* Does the model we have created accurately mimic the real world? Have the relationships and factors that we have sought to demonstrate with the model been demonstrated? How like reality is the model? Is it very accurate? Too accurate? Not sufficiently accurate? Inspection of the model and comparison with the "real thing" allows us to fine-tune and make last-minute adjustments to bring the model into line with our concept of what it should do, be, and indicate. The comparison can be a simple inspection, as with an icon such as the ship model, or can involve one or more trial runs to test a dynamic model's response to motive forces, as with a glider, or, for that matter, the economic model. The ship need merely *look* like the real thing. The market model must *behave* properly in reaction to market changes and *be predictive* of change that might occur in the future. And that means testing the mechanization as well as the format. A well-formatted model that does not behave as it should is useless for purposes of predicting behavior. Anyone who has spent the slightest amount of time building computer programs is well aware of *that* reality, to be sure.

6. *Adjust the model as necessary to achieve the desired "fit."* Final adjustment in accordance with the comparative inspection of the fifth step is the last action in building a model. It represents the finishing touches that a modeler puts on the model, since all that could be done has already been done by the modeler to achieve accuracy and purpose in the construction of the model. It is the last step, but in a changing, dynamic world, where the interrelationships among factors of society and technology are constantly rearranging themselves, it is a never-ending step, involving the modeler in a constant reassessment of the model, its usefulness, and its ability to describe the real world. In the future, the model may change dramatically, or it may remain stable, or, what is most likely, it may simply be replaced with a better, more

accurate, more useful model. Models are never static if they have accuracy, because the world that they are designed to describe is dynamic, ever changing to offer new challenges and new information to be gained from its investigation.

MATHEMATICAL SIMPLICITY

Before leaving the general subject of models to delve into two specific forms of modeling useful to the investigator of social and technological structures, a brief explanation should be made of the general class of models known as *mathematical models.*

Mathematical models are considered to be a type of symbolic model. Symbolic models deal with ideas and abstract approaches rather than physical or mental constructs. They are descriptive and predictive as are other useful models, and the particular importance of the mathematical model form of the symbolic category is that it can succinctly and briefly describe extremely complicated real-world relationships in very little space. Unlike the verbal type of model, also a symbolic form, the mathematical model is not easily misunderstood. Mathematical relationships are very tightly defined, and their interaction cannot easily be mistaken. Any given formula will always describe the same relationship among factors, the only difference being one of magnitude, as represented by the variable quantities that can be plugged in. The formula $Y = 3X + 7$ will always result in an upward-sloping curve when plotted on a Cartesian coordinate system. It can do nothing else, no matter what real (positive) number quantities are entered for the independent variable, X. There can be no confusion. Every graph of this function will be identical in plot to every other, by definition. This is not always true of verbal models. Mathematics is exact and is therefore valuable as a diagnostic tool, since we can be certain of unvarying answers to formulas, if the information we seek can be quantized. A reduction of a picture to mathematics, as with a computer scan of a lunar landscape or the Mona Lisa, ensures absolute identical reproductions if translated into the original medium. Verbal models such as descriptions, on the other hand, are vague, inexact, and, by their nature, subject to the interpretation of the observer. Try asking four people to describe a building, and compare that with the results of asking four scientists to describe mathematically the results of dropping a ten-gram weight from a height of sixteen meters in a vacuum, measuring the height from a reference point of sea level. The difference in the models and their exactness becomes quite apparent.

Mathematical models are classified in accordance with a series of dichotomies that define their characteristics. They are said to be either *quantitative* or *qualitative, probabilistic* or *deterministic, general* or *custom constructed, descriptive* or *optimum.* Briefly, the dichotomies define the mathematical models as follows.

A model is either quantitative or qualitative depending on the ability of the user to accurately define information numerically. In the case of *quantitative models,* the measurements used to determine the values of dependent and independent variables in the formulas are accurate and discrete. I can count the number of seconds it takes a ball to drop sixteen meters with a very high degree of accuracy. The result of an experiment to do so would yield a specific, discrete value for elapsed time. This would be a quantifiable measurement and fit nicely into a quantitative model. As an example, consider the formula, $P = TR - TC$. This formula describes the profit received from the sale of goods as the difference between total revenues (TR) and total cost (TC). All that is necessary to find a definite value for profit is to know the total amount of money received for the sale of goods and the total amount of money paid out in producing those goods. The difference is profit, and that had best be a definite, discrete number. (If it is not, the IRS will be most happy to assist the businessperson in finding out why.)

Qualitative models are not exact. A *qualitative model* is a method of encoding inexact concepts numerically. This is done through the use of a branch of mathematics dealing with continuous rather than discrete functions known as *statistics.* Qualitative mathematical models indicate the *general* results of behavior rather than the exact results. The difference lies only in the ability to accurately measure the phenomenon under study. An excellent example of a qualitative model is the use of a survey to determine public opinion on nonmathematical issues. By asking a group of people (sometimes referred to as a "statistically significant population," meaning a large enough sample of opinions to represent a general attitude among the general population) to rank in order of importance a series of ten characteristics that they would like to have in a spouse, it would be possible to determine to some degree the types of traits that people look for in potential mates. The results would not be equally true for all participants, and there is no way of physically measuring the importance of one trait over another, yet the attitudes and expected behavior of a population can still be predicted as a result of the survey. Qualitative models are growing in importance, particularly with the advent of the computer and advanced statistical methods that allow the investigator to process huge amounts of data and test models for accuracy. This is a growing area of modeling that holds great promise for the future.

In a similar manner, mathematical models can be considered either deterministic or probabilistic. *Deterministic models* are formulas that always result in the same, precise answers, whereas *probabilistic models* express tendencies. The probabilistic models use such concepts as the mathematical mean and standard deviation to describe accurately what the *tendencies* of a system are, or to develop probabilistic formulas based on repetitions of trials to calculate "the odds" of something happening under a given set of circumstances. An example would be determining what percent of the population is viewing the first half hour of the 7 o'clock news over the NBC network affiliate in San Bernardino, California. We could conceivably go around to every household on a given night and ask, but the likelihood of success and the expense of achieving that success would probably far outweigh the benefit we would receive from the information. And that would only tell us how many people watched the broadcast at that time over that channel on that one night. What if we wanted to know the figure for the entire year of 1985? Accuracy in this case would be expensive and impossible from a practical standpoint. Does the modeler simply give up and decide it can't be done? Not if he or she works for a marketing firm and wants to continue to work there. The modeler *estimates* with as high a degree of accuracy as possible and then indicates how much confidence he or she has in the estimate. Chances are the exact number is not even a useful thing to know. The trend or tendency of people to behave in a given way is more usable and more important.

In the case of the trial-and-error approach to probabilistic models, consider the local weather report. What is presented as a predictive forecast is actually a report of the *odds* of certain weather patterns developing in the future based on an analysis of what has happened in the past when conditions were similar or identical to current conditions. No weather forecaster is rash enough to tell his or her audience what *will* occur tomorrow, only what is *expected* to occur.

With a deterministic model this problem of probabilities does not exist. *Deterministic models* are so accurate that they indicate exactly what to expect in a given situation. Deterministic models predict measurable phenomena, such as force of impact, speed, or the number of electron volts created. The results of these models are predetermined and exact rather than relative.

One caveat is in order concerning deterministic models. Technically, they are also estimates of true quantities, albeit highly accurate ones. The exactness of the model (the degree to which it is deterministic) is relative in nature, being only more or less accurate than some other system of measurement. A foot is exactly twelve inches and that is a deterministic quantity, yet the measurement of the length of a foot or even an inch is something that becomes more and more exact with improvements in the technology of measurement. At present, the measurement can be made with laser beam technology to such a high degree of accuracy that it is useful only for a very limited portion of the population. For most of us, a ruler or yardstick does just fine, inexact as they are.

The dichotomy between ready-made and custom models indicates whether the model is a normal mathematical relationship that is generally accepted for a wide number of applications or a model specifically designed to deal with a singular phenomenon faced by the modeler. Many custom models contain ready-made models as part of their structure. By way of example, the formula, $V^p = V^t/(1 + r)^t$ is a standard model form that allows anyone to calculate the present value of money received in the future after allowing for the time differential involved. It is a very useful model that is used in business, banking, finance, and other money-oriented disciplines. It is a ready-made model of the idea of present value. In contrast, a company may wish to create a custom model of the inventory system used by the company as part of an attempt to computerize routine operations. Such a model is based on the inventory system and model needs of the one company. It would be *customized* for that company's individual situation, though it may contain a large number of standard functional models used in inventory control.

In the case of descriptive models versus optimizing models, we deal with a difference in purpose. *Descriptive models* have as their purpose to do exactly what it sounds like they should do, describe. They are mathematical representations of real-world phenomena and nothing more. The *optimizing model* is distinguished from the descriptive model in that it seeks to find the optimum combination of actions to create a desired result. The optimizing model compares combinations of variable inputs in order to find the maximum, minimum, or most acceptable type of result for the purposes of the modeler. Determining the most efficient speed for an automobile through modeling is an example of this. There is a trade-off between speed and fuel consumption for most automobiles, and this conflicts with what we would like to have, that is, minimum fuel expense and maximum speed. The two are simply not possible together. However, by calculating how fast fuel consumption changes with a change in speed, it is possible to find the *optimum* or most efficient combination that is acceptable. How much reduction in speed am I willing to trade for an increase in fuel economy? It is this sort of optimizing analysis that leads to decisions on speed limits. Other models of this type can be used to determine the size of a manufacturing plant, the design for city plans, the size and shape of office buildings, the use of materials, the horsepower requirements of an automobile, and a wealth of other expense-benefit–type considerations.

GAMING AND SIMULATION

At first inspection, the terms *gaming* and *simulation* appear to have identical meanings. Yet as with any closely associated terms, the fact that there are two distinctly different words available to describe the same primary process indicates a difference, a shade of meaning, that separates the two. In essence, simulation can be considered to be a very specialized form of the game. It is because of the specific characteristics of the simulation that it is particularly useful in studying the interactive systemic nature of society and technology.

Games are considered to be any recreation or sport incorporating specific rules that require the participant to compete in some way, either against other players or against himself or herself in attempting to achieve some specified goal. Note that in order for an exercise to be a game there are rules to be followed, a goal or winning condition to strive for, and competition against one's self, as in comparing past scores with present scores, or competition against others. Competing against the game itself, as in solitaire or computer games, can be considered competition against one's self. Even team games whose object is to optimize scores against the game itself can be considered self-competition. The level of skill one has or one's team has is an indication of improvement in playing the game.

Note further that a game is considered to be a recreation or, more exactly, a re-creation, that is, something that recreates a condition or set of circumstances reflective of a real-world phenomenon. Therefore, even formal games are obvious models.

Games can be loosely classified as games of chance (poker, roulette), games of skill (competitive sports, chess), or a combination of both (bridge, "Risk," war games, and so forth). They are also often classified according to the type of equipment and competition involved in their play (card games, contact sports, board games, computer games, and so forth). All of this may be interesting to the inveterate gamer, but somewhat academic for our purposes, with the exception of a particular type of skill-chance-oriented game—the simulation.

A *simulation* is a model that copies the behavior of some aspect or aspects of reality. Simulations attempt to describe and test behavior patterns of interactive systems and then to be predictive and descriptive of changes in the system that may result from changes in one or more specific elements (subsystems) within that system. The creator–player is interested in recreating real-world conditions on some limited basis for the purpose of aping (mimicking) the behavior of the real-world system and then inputting changes to gain skill in dealing with the results of environmental change. Certain characteristics of simulations make them specific to their form:

1. Simulations deal with real-world systems with a level of detail ranging from the simple to the quite sophisticated.

2. Simulations are interested in the interactions of subsystems that exist in the overall system that is being studied.

3. Simulations are designed to mimic real-world behavior, not only in accordance with known relationships but also the innate uncertainties of the real world by including reasonable uncertainties in the form of chance variances in output as a result of given inputs.

4. Simulations are system specific, dealing with individual problems or scenarios to allow the participant to gain experience artificially in interacting with real-world conditions.

5. Simulations have the capacity to predict the results of many different combinations of conditions in rapid succession, allowing modelers the opportunity to utilize the strengths of trial-and-error decision making without many of the shortcomings.

6. Simulations are both quantitative and qualitative, offering probable outcomes to given actions.

Not all of these characteristics are obvious in all simulations, but they are nonetheless present.

Examples of the use of simulations are present everywhere. Board games such as "Monopoly" and "Risk" are obvious examples of scenario simulations, dealing with real estate markets and world military domination, respectively. Computer arcade games are often simulations, offering the participants the opportunity to test their skills against electronic foes in everything from space warfare to guiding supertankers through the coastal straits of Alaska. Flight trainers are highly sophisticated simulations utilizing the analog model to give pilots the feel of an aircraft and allow them to work their way through all of the possible hazards that they might encounter without risking life and limb. In business, simulations are utilized to sharpen management and marketing skills, to outthink the competition, to study the effects of proposed strategies, and to test products. Engineers use computer and analog simulations to design and test equipment, machinery, and hardware. The military uses simulations to practice the art of war, often quite realistically. There are even computer simulations designed to simulate the operations of other computers!

SIMULATION CONSTRUCTION AND DESIGN

Designing a useful simulation is essentially the same as designing a model. The differences lie in the purpose of the simulation. Briefly, the process is as follows:

1. *Define the problem.* This may seem self-evident, but it can cause a great deal of consternation in later steps of the process if not properly done in the beginning. People often find themselves hampered by not completely understanding what it is that they are trying to do. A poor definition of the problem results in a severely limited understanding of how to solve it. The modeler may end up simulating the wrong system or not developing true relationships. The way a problem is defined dictates what solutions will and will not be sought.

2. *Conceptualize the system.* This is nothing more than a restatement of the mental model building used in initially conceiving a model's characteristics.

3. *Create a model representation.* Build a primary "rough draft" of the simulation as conceptualized to find the hidden factors that have not yet been considered.

4. *Observe the model's behavior.* Find out if the initial representation behaves as was initially expected.

5. *Evaluate the model's behavior.* Do you need to alter your conception of the problem? Do you need to adjust the factors involved in the simulation? Are different elements more important than those you initially took into account? Are the results consistent with your initial purpose?

6. *Adjust the model as necessary.* As Alcorn's corollary to Murphy's law states, any computer program that runs the first time is either worthless or you've missed something. The same is true of simulations.

7. *Use the simulation.* Test it to discover how closely it reflects reality. Some of the most useful simulations have been the ones that did *not* mimic real-world events and yielded information about contributing factors of major importance that were *not* included in their construction. Just as a well-formed experiment can be as valuable if it fails as if it succeeds, so the simulation can be useful in *not* initially working. Constant adjustment of criteria will, in all likelihood, be necessary in the case of complicated computer simulations, if they are expected to ape the activities of a dynamic, real-world system.

CAUSE AND EFFECT: ONE, TWO, THREE

Interactive real-world systems demonstrate cause-and-effect relationships among their subsystems, as do the sub-subsystems within each subsystem. From the overall to the smallest subset of a system, cause-

and-effect relationships link elements together in behavioral relation-ships. In simulations, these cause-and-effect relationships can be used to organize activities cohesively. It is a prime method of construction to approach systemic simulation from this viewpoint.

In creating a simulation model, the patterns of behavioral relation-ships indicated by cause-and-effect interactions fall into a natural hier-archy. Since cause-and-effect relationships are seldom isolated, each reaction (effect) resulting from some initial causative action is in itself a causative action leading to some other reaction. The occurrence of rain, as an example, could be the cause of plant growth, the plant growth being a primary reaction to the increase in available water. However, the plant growth triggers other activity, such as an increase in parasitic insects in a farmer's field that only occurs because of the presence of the plants. This is a *secondary effect* of the rainfall. And as a result of the increase in the insect population, the population of insect-eating birds may rise; this is a third-level or *tertiary effect.* This multiplicative expansion of cause and effect from primary to secondary to tertiary illustrates the *chaining effect* in real-world systems.

All systems are related to all other systems. They are all part of the same process, all interconnected, and all quite interactive. Because of this, the modeler must decide which elements to include in the simu-lation and which to leave out, based on the aims of the model. Like-wise, there are times when important secondary or tertiary effects may exist as a result of a given action, yet they may pass unnoticed because of the limits of construction of the model. Simulations will differ from reality in accordance with this fact.

It is additionally important to consider the concept of *feedback loops* in the construction of simulation models. These loops describe the interactive nature of certain give-and-take interactions of systemic subsets among themselves. Loops feed on themselves, so to speak, by each element contributing to the functioning of the other in some way. As an illustration, consider the relationship that exists between educa-tion and wealth. It could be argued that the occurrence of wealth and the level of education in a population tend to be mutually supportive. The greater the wealth of a culture, the greater its capacity to devote time and effort (and money) to educating that population. Similarly, the greater the level of education, the more productive and therefore the more wealthy the population as a whole can be expected to be. It would seem that these two factors are indeed mutually supportive. This is what is known as a *positive loop.*

An equally valid case could be argued for the relationship between farming and soil erosion. As farming increases in an ecology restricted by finite farmland, the intensity with which that farming takes place rises, resulting in overuse of the soil and eventual erosion. The erosion results in poorer crops, which necessarily means more intensive farming to keep up with the demands of the population for foodstuffs, resulting in more erosion. More farming produces more erosion, and more erosion produces more intensive farming. This is a *negative loop*, a situation in which each subsystem negatively affects the other subsystem, thus resulting in a net downward spiral of conditions.

Both positive and negative feedback loops can be found in real-world situations. The job for the modeler is to determine which loops are most important, how they affect the overall system being modeled, and how they should be taken into account in building the model. The modeler must also keep in mind that each subsystem is in itself a collection of loops, resulting in secondary and tertiary effects to take into consideration.

To illustrate the simulation technique, the following deals with the development of a simple model to explain some relationships that exist in a real-world ecosystem. For our illustration we will modify the farming example. As a primary cycle, we will use the relationship between farming and population. With increases in farming, the availability of food initially rises, resulting in more food and better health for the population. A healthier, better-fed population is economically successful, and, as with all successful economic systems, it grows as a result of the success, and its population rises. The rise in population results in the need for more food, which means more farming. Here we may consider the primary cycle to be positive and self-supporting in that farming creates more population and more population creates more farming. The loop is represented diagrammatically in Figure 7-1. The arrows in the figure indicate the direction of the cause-and-effect cycle to remind us of what is triggering what. The use of the plus sign in the center indicates the nature of the loop, in this case, positive.

If we stopped at this point with our model, we would find a constant increase in farming and in population through time. And yet we know that this is not a very true picture. Too many secondary and tertiary effects have been left out of the analysis. There are too many

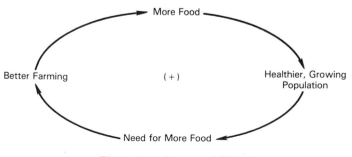

Figure 7-1 Cause and Effect

other relationships to be dealt with before even the simplest degree of accuracy can be obtained. For further clarity, a closer examination of subsystem elements should be investigated as well. One of these we have already shown in illustrating the negative effects of farming on the erosion of soil. Figure 7-2 illustrates this erosion-farming mechanism.

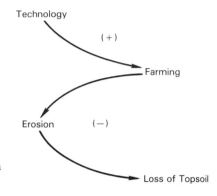

Figure 7-2 Technology and Erosion in the Farming Cycle Model

The loop in Figure 7-2 is negative, indicating the dilatory effects of soil erosion on increasing farm output. Erosion acts as a governor, slowing down the farming process as more and more effort is put into intensive farmland use. These secondary effects usually present themselves in the real world and are easily taken into account. They also help to explain other secondary effects that may arise in connection with the primary cycle. As an example, it has already been indicated that technology is a method by which humans control nature through the production of artificial constructs to counteract negative pressures in their lives. If people are cold, they learn to build fires. If they are hunted by beasts, they learn to defend themselves and become the hunter. It is equally true that if farming becomes difficult, technology is developed to improve it. The reduction in farm output due to erosion may well be counteracted by an increase in farm technology, creating the opportunity to produce more crops on less land or, alternately, to reduce erosion. Taking this aspect into account yields still another loop by which erosion and farming needs create technology. This is represented in Figure 7-2.

The discovery and inclusion of secondary effects in a simulation model pose little difficulty. Tertiary effects and possibly quaternary effects and beyond can be more of a problem. It is often the tertiary or quaternary effects that prove to be the most important in analyzing a real-world system, not necessarily because of their impact per se, but rather because of not being considered in the analysis. It is difficult to react to a factor that has not been considered.

A case in point would be the result of soil erosion and the loss of topsoil on other elements of the ecosystem. If erosion takes place and topsoil washes away, that soil must go somewhere, often into rivers and lakes that silt up as a result of the increase in dissolved solids in the water. The lakes become muddy and clogged, and rivers become unnavigable. Fish populations decrease as oxygen supplies in the water decrease, and aquatic vegetation thins. Local industries are hard pressed to dispose of waste once easily handled by the ambient water systems. Stagnation may result. Mosquito populations and possibly disease are on the increase as a result of the process. These effects may be the most serious and least understood of all the consequences of the initial increase in population and farming. It is this type of effect, the hidden, not immediately yet potentially dangerous consequence of some system change, that is most necessarily understood. A complete model of the simplistic system appears in Figure 7-3.

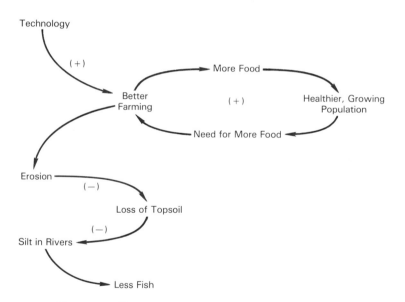

Figure 7-3 Farming Cycle Model—An Extended View

How does the simulation technique help us in our study of technology and society? It is exactly for this type of complex, multilevel system that the simulation technique was devised. It is a method of using the systems technique of defining and subdefining a real-world structure into an understandable network of cause-and-effect behavior and influence, so that we can both comprehend the nature of the problems of technology and society and predict the effects of changes before they take place. A further refinement of the process is to quantify as many of these relationships as possible to develop the predictions. The exercises at the end of the chapter offer an opportunity for students to gain firsthand experience in using simulation to enhance their own understanding of systems and the manner in which they operate.

CONCLUSION

In pursuit of understanding the physical world and the various ways in which its components interact, humanity has developed the facility to create artificial constructs that behave in similar ways to observed patterns in the physical world. These constructs are known as models. They take several forms, depending on the nature of the phenomenon being studied and the characteristics that the individual investigator desires to understand. That is, the model's form is determined by the limitations of the environment and the nature of the individual's questions about the real-world phenomenon in question.

Models can be usefully classified as *analog*, a type that behaves in some way similar to the reality that it is designed to represent; *iconic*, which is designed to resemble a physical reality though not necessarily to behave in an analogous manner; *verbal*, which is a model designed to convert thoughts or concepts into language, establish relationships and restrictions of real-world systems and organize them in an understandable form; and *mathematical*, a symbolic manipulative representation of reality designed to describe relationships among certain factors of the reality that it is designed to represent.

Of special interest to the investigator of techno-sociological relationships are simulations and games. These forms of models may involve any or all of the above-mentioned types of models. Games are chiefly useful in creating experience artificially by providing a competitive environment in which an investigator may mimic interaction with a particular environment based on assumed rules or parameters used to describe the environment. Simulations provide a construct for describing and predicting behavior in real-world situations within the confines of specific parameters. By the use of gaming techniques and simulations, an individual is able to develop expertise in dealing with complex systems and in predicting outcomes before a commitment to some line of action is finalized. In addition, the very construction of games and simulations forces the individual investigator to discover the nature of the relationships of the real-world phenomena in question.

Models in general and games and simulations in particular offer organization, structure, cause-and-effect relational concentration and predictability so important in the study of techno-sociological systems.

THOUGHT AND PROCESS

1. In light of the information provided in this chapter, return to the questions at the conclusion of the last chapter and try your hand one more time at the process of using the systems analysis technique.

2. The world is filled with models. We encounter scores of them each day. Look around and find an example of each type of model discussed in the text among the common experiences of your daily life. Why do you suppose the particular form of model chosen by the modeler was preferred in each case over the others? What do you believe his or her purpose was as indicated by the form chosen?

3. Design a board game based on one of the following themes. The thematic titles should be self-explanatory. Let your creative instincts run free. (a) A Date with a Childhood Sweetheart, (b) The Battle of Atlanta, (c) Stock Market Tycoon, (d) Space Travel Adventure, (e) College Registration Survival, (f) Invasion of the Mechanical Man, (g) Zombie's Feast, and (h) Gridiron Glory.

4. Using the modeling process outlined in the chapter, develop a model for some real-world phenomenon in which you are interested. Go through the process and document your choices of form, method, and so forth.

5. Build a diagrammatic simulation model of the various cause-and-effect relationships that are included in some real-world phenomenon in which you are personally interested. Carry your analysis to at least the third level of effects (tertiary-effect loops).

CHAPTER 8

The Systemic Model

The following discussion centers on a list of questions designed to be a systemic outline for the study of technology and social systems. It has been created using the basic methodology discussed earlier and is presented as an outline of study for any technological change or group of changes. The questions asked and the categories chosen are designed to survey possible areas of social structure that may be affected by changes in technology as they occur. Using the outline approach, it is hoped that the student will be able to focus on the most likely future effects of any given technology.

However, a caveat is in order. As with any discussion of social structures, particularly speculative discussions of future conditions, the one thing of which we may be certain is that no matter what we discover of the probable effects of technological change on the culture through the manipulation of this set of categories, what actually happens will be somewhat different from what we predict. There is no method available to the investigator, no matter how skilled he or she is, that allows the investigator to be absolutely accurate in prediction. Yet by predicting, manipulating social factors, speculating, and then comparing speculations with reality, the investigator improves the quality of the predictions made. Only through experiencing the "art" of futuristic prediction can an investigator be able to improve his or her accuracy.

Please note that the factors to be considered are subsets of the cultural whole. By applying this method to historical subject matter, that is, by investigating how extant technologies have related to the culture, we gain insight into what should be expected in the future. What does it matter if we are not absolutely correct? Our purpose is not to find a specific definitive answer that does not exist. It is to improve our ability to deal with the future. If we can improve our accuracy, we have gone a long way toward understanding our society, our technology, and the place of both in the larger system that is our world. Even physics, one of the most definitive and precise sciences, is quick to admit that it deals with only probabilities in the physical world. Who are we to attempt more with something as nebulous and multifaceted as the sum total of experience of the entire human race?

A brief discussion follows of each of the elements in the system that are deemed to be most important in studying technological and sociological change. It is hoped that it will be equally applicable in the study of any facet thereof.

AREAS OF INTERACTION BETWEEN SOCIETY AND TECHNOLOGY

Technology's Effect on Commerce

The first field deals with the economic and commercial consequences of the technology in question, whether it is an existing historical technology or a burgeoning one that is just beginning to impact the society. This is not chosen without considerable thought. The commercial aspects of technology have already been extensively covered in Chapters 4 and 5. By this time, it should be apparent to the student that the impact on this aspect of human culture is a major one.

Most technological changes begin in the economic realm. Technology is a key factor in the supporting and developing of an economy, in the securing and maintaining of jobs for the population, and most certainly in determining the level of economic welfare experienced by the members of the society. What is the effect of the new technology on business and commerce? Does it represent new goods and services? Are we dealing with new products resulting from technological change? If so, how will the new products impact the economic structure?

One source of new technology is the search for increased economic efficiency and a sincere desire to reduce the cost that the society has to pay for the availability of goods. Whether the purpose of the technology is to improve its effective use of available natural resources or to increase or alter the supply of available resources, it will impact what people buy, what they choose to do with their time, what jobs are lost due to changes in the overall economic mix, and possibly the price of other goods that compete with the technologically changed function for raw materials, labor, and the limited capital resources that are available.

As a secondary effect, how will the new technology create changes in unrelated or distantly related markets? The computer was a fantastic new technology. Indeed, it is still so. The changes that have taken place in the business world reach far beyond the immediate impact anticipated. On the surface, it was not difficult to realize that the computer would affect the market for mechanical devices designed to do tasks that a computer does, such as adding machines, typewriters, or even automatic mechanical control mechanisms. Yet this consideration does not begin to deal with the other changes that have taken place in the business world as a result of computer technology. An entirely new industry had been born in the form of the microcomputer, a stepchild of the larger computer industry available only to big businesses with big dollars to buy big computing power. New skills and new opportunities for employment have come about as a result. The expansion of the computer market and proliferation of microcomputers into the mainstream of American life have increased the ability of the homemaker to run a household effectively; to shop and cook more efficiently; to learn about the true nature of family finances; and to create more leisure time for spending money, for education, for hobbies, for part-time productive jobs, and for a host of other things. The nation has, in effect, stepped out of the industrial arena into the information arena. Here is a single technology that has so changed our ability to gather, store, manipulate, and disseminate information that the entire economic structure has been transformed. And all from a single technological change, albeit with a highly sophisticated and extensively proliferated collection of applications. The computer and its impact on the economic structure is no less startling and dramatic than that of the steam engine; electric power; or, for that matter, fire and the wheel. Through the computer, the efficiency and expansion of the economy have been so greatly accelerated that we are hard pressed to keep up with the changes. It is indeed revolutionary in nature and explosive in the speed at which it is altering our economic lives. Answer two questions: Ten years ago, how many people did you personally know who had access to a computer or dealt with one? How many do you know now? If you answer honestly, you will be astounded at the manifold changes that have taken place in just a single decade. And the next decade promises to outstrip the last by far.

Technology's Effect on Social Systems

The extent to which a technology affects social systems has been briefly discussed throughout this book. Specifically, the emphasis of this question deals with basic patterns among social groups and the changing patterns of needs and need fulfillment resulting from technological changes.

When the Industrial Revolution came about and particularly when the industrialization of America took place in the last century and first half of the present century, a number of social factors changed, which would impact not only those directly involved in the process, but those who chose not to be involved in the industrial boom.

Workers in an industrial setting are able to command higher wages than farm workers. This is a fact of economic life. It is the result of the efficiency of labor in an industrial setting compared with the efficiency and productivity of farm workers, on which, at the time of the first industrializing moves in America, the country's economy was based. Economic systems recompense workers in accordance with their productivity rather than how hard or how long they work. It is their production level that determines how valuable they are. For the industrial worker, whose level of productivity working in a mill or production plant is as much as ten times what his or her farmer counterpart could achieve, this meant ten times the wages for the same amount of work by simply shifting from farming to industrial work. Thus the mass migration from the country to the city occurred, with its accompanying rapid rise in urban population.

The industries were located in urban areas near supplies of raw materials, centers of transportation and communication, and markets. Chicago rose as the center of the meatpacking industry. New York grew through its transportation, communication, and financial centers, as did Houston and Atlanta at a later time. Pittsburgh, in the center of the coal and iron ore belt, was a center for foundries as early as the 1850s. Hundreds of other examples support the same argument. Industrial concentration and bigness, being the way to achieve efficiency, meant concentrating industry in small areas, which led to the packing of population in and around those locations where high-paying, high-productivity jobs existed.

With these heavy concentrations of large numbers of people came all of the attendant problems of urban life heretofore no more than mere annoyances to the general population. Families lived closer to one another. The unavailability of living space created multistory and multifamily living. Families found themselves in close proximity to neighbors, unable to depend on themselves for food and simple tools, dependent instead on supplies bought in local neighborhood markets. With crowding came an increase in crime, an increase in disease, and an increase in stress on the family unit.

City life, with its compacted physical structure and rapid pace, replaced the easygoing, steady pace of rural communities. One no longer knew everyone in the neighborhood or in the community. People came and went more frequently. Mobility increased for some and decreased for others. Time telescoped as efficiency in business spread as a concept to efficiency in life style. Dispersion, particularly among generations, tended to decrease the level of interaction among members of the extended family and to increasingly isolate the primary family unit.

New social institutions have arisen as a result of the industrialization process. Unions arose as groups of workers fought for their collective rights. New groups within the work environment have risen to satisfy or thwart many of the needs of individual workers who are no longer able to obtain need fulfillment through the extended family. Identification with cliques, social groups from the work environment, or the company itself forms social structures for the benefit of the urbanized worker, a condition neither necessary nor possible in the older, agrarian culture of preindustrialized America.

These processes are still taking place in the late twentieth century as our technology alters our perceptions and our patterns of living. Television and telephone communication have replaced the more personalized forms of communication, again isolating us from one another and reducing the opportunity for interaction and the need to form social structures through traditional channels. The high-tech high-touch concept of John Naisbitt is no myth. It exists and promises to have a strong influence on our social behavior in the foreseeable future.

The key is to determine how a technology will affect the *opportunity* and the *probable form* of social systems as it alters our day-to-day lives. As an evolving species, humans should expect change, and the changes in social structure that result from technological innovation will determine to a large degree the quality and kind of life available to us in the future.

Technology's Effect on the Environment

Technology's effect on the environment has received much attention in recent years, mainly due to (a) the reduction in lead time perceived between the instigation of a new technology and the serious effects on the ecological balance of our world that may possibly occur as a result of it, and (b) the greatly increased control over the environment that modern technology represents.

Environmental considerations are among the most obvious areas of importance to us in dealing with technology because they are utmost in our minds. Every technology affects the environment to some extent, just as it affects every physical entity to some degree. Many of the technological advances brought forth in recent years have been viewed as detrimental to ecological balance. Acid rain threatens wide ranges of forests and farmland. Pollution from nuclear tests is suspected of causing cancer in victims unfortunate enough to be exposed to it, while chemical content in rivers and streams destroys communities and reduces the productivity of already overworked soil. Diversion of water from natural sources feeds towns and cities, only to create shortages elsewhere. Leisure use of natural habitats destroys landscapes and threatens the homes of wildlife. Pipelines are purported to damage the ecologically delicate balance of permafrost environments and to disrupt the migration patterns of elk. Oil spills pollute our oceans, the main source of oxygen for the planet, leaving a trail of tar and oil solids from one continent to another. The list could go on and on . . .

Yet it would be inequitable to consider only the negative impacts of technology on the environment, though they may be a serious matter of concern. In addition to the detrimental effects that technology can be viewed as creating, there are also positive effects. It is through our understanding of nature and the manipulation of its laws that we can construct dams and spillways to bring life to the deserts of the Near East. It is through our understanding of nature that we can prevent disaster by destroying diseases detrimental to wildlife or save endangered species facing starvation and extinction, not from the hands of humankind but from the pressure of natural droughts. Science and technology can reclaim natural wilderness areas as well as destroy them, protect the integrity of ecological systems as well as disrupt them, and prevent catastrophic occurrences as well as create them. As with any other human system, it is the use to which technology is put that determines its desirability, not the nature of the technology itself.

Technology's Effect on Individual Psychology

Our attitudes, opinions, approaches to problem solving, and psychological balance are all affected by changes in technology. The world in which we live, including technological change, includes all the inputs we use in developing our personalities. Experiences teach us what to believe about the constitution of our world. Observations shape attitudes about social interaction and what is and is not considered appropriate. Threats from external sources create the need for a host of adjustment mechanisms, which collectively add up to a considerable part of our behavior patterns.

The vignette in Chapter 2 describing a hypothetical meeting of Luddites illustrated the pressures of industrialization in early nineteenth-century England as the cause of a mass movement. Violent action resulted from the inability of the members of the movement to cope with the changes that were taking place around them. This is an example of how technological change can affect the otherwise stable thinking patterns of a human being.

In modern times, the degree of specialization and separation of workers from their tools inherent in the "big business" approach to industry results in feelings of alienation among workers. Personal satisfaction of needs drops as workers become less and less attached to the finished product with which they are working. There is no feeling of accomplishment, no personal interaction, and therefore little identification with either work or company. Patterns of behavior that would be considered unthinkable in a more personal form of endeavor become commonplace, including a reduction in pride, a reduction in honesty, and an increasing dissatisfaction with work as a whole.

There are also the feelings of frustration, fear, anxiety over the unknown, and dislike toward others that result from the interaction of the individual with the technologically changing society. With the proliferation of television sets in America, the methods by which the population internalizes information changes. The presentation of the Vietnam conflict in detail and, at times, within hours of actual occurrences, is a case in point. For the first time it was possible to sit in one's living room and be a part of the carnage, the fear, the death. Such a strong input was beyond the capacity of many in the society to cope with, and the result was a peace movement that eventually brought about an end to U.S. involvement in that war. Countercultural elements, whether representing a peace movement seeking an end to carnage or merely malcontents not willing to interact with a world that is changing too fast for them to internalize, are the result of changes in psychology or the result of an inability in people to change attitudes in the face of changing times.

How will a given technology change the psychology of a nation or a given group of people within a culture? How will a society react to a new methodology? Will it ignore it? Will certain people attack it as the Luddites did on the basis that it is "stealing" jobs? What about society's attitudes concerning technological advances in the past? Will a person who grew up on science-fiction movies be predisposed to fear robots? Do people view new technology with suspicion because of the Frankenstein and Dr. Jekyll stereotypes? How do they handle the creation of artificial life? What has the death threat of nuclear war done to their attitudes and ideals? Will future technology have a similar effect or will it act to ease tension, alleviate our anxieties, and issue in a new era of confidence? These are the kinds of questions that need to be asked concerning technology and individual psychology.

Technology's Effect on the Rate of Change

Again the prime example of this issue in the modern world is the computer. Never have we been able to manipulate and have available to us such huge stores of information. With this availability comes a

heightened possibility for progress and development of still other technology. With a sharing of knowledge comes a sharing of wealth that can
only create more wealth through its use.

But this is certainly not the only example that can be cited. The
oared galley, the railroad, the automobile, and the airplane are all
examples of technological changes that brought about a flowering of
progress in individual societies. In the case of oared galleys, the ancient
vessels so equipped were instrumental in initiating international trade
by opening up foreign lands in the ancient world as they made their
way across oceans and seas, disseminating knowledge and spreading
local technologies to other lands. Until the railroad opened up the West,
the movement of population outward from the coastal areas was extremely slow in the United States. Railroads meant faster, safer travel.
They were a means of expanding markets by shipping greater amounts
of goods from one area to another. Markets were no longer local. Per-
ishables could be sold over long distances. The movement of people and
raw materials decreased dependence on limited, highly favorable habitats. The automobile opened up the entire country to the population,
connecting towns and villages with cities. Whole new industries sprang
up, creating still other industries in an orgy of growth that did not slow
down until midway through the twentieth century. The airplane meant
fast travel for passengers. This was accompanied by increases in the efficient use of time, particularly in business, and in the growth of holi-
day travel, which caused still other industries to flourish. In each of
these cases, the speed with which progress was made was magnified. By
facilitating the spread of technology, whole societal structures were
created.

Technology's Effect on Institutions

Technology reshapes a society. In that capacity, it must necessarily impact on the social institutions of the society. Social institutions
are like other social constructs—they are in existence because they fulfill some purpose for the population. If the introduction of a technology into a society changes the needs of the population or alters the
availability of the institution to perform the function, then the institution itself is affected, either by altering its form or by disappearing
altogether. Some examples will clarify this point.

Religion is one of humanity's major institutions. Yet it has often
clashed with changes in the structure of society. During the Renaissance, there was a reevaluation of the concept people held of the world
and an expansion of understanding that effectively destroyed the old,
traditional paradigm of Western Europe forever. The Church, as a political, religious, and administrative source, found itself in the thick of
battle over many of the new technological and scientific discoveries

being made. Galileo utilized the telescope, a technological device, to study the planets. For doing so, he nearly lost his life at the outrage of Church officials. Copernicus put forth the idea that the earth was not at the center of the universe, an idea in direct opposition with the Church-held view of the time. He was threatened with excommunication and being burned at the stake as a heretic unless he recanted those views. Martin Luther awoke to the inconsistencies within the Church hierarchy and created a new approach to Christianity, called the Reformation, which was the result of increased knowledge of the world through reading books, a technologically oriented item, that had been printed on printing presses, technological devices made by artisans using technological techniques of the day.

Indeed, the practical, worldly nature of the Protestant ethic—the moral code that so inspired the founders of this country and helped create the mercantile empires of Europe—had as its impetus the technologically fruitful age—the Industrial Revolution. Religion tends to change as the consciousness of people changes. And technology can do much to change people's consciousness. Religion also benefits from technology. The spreading of the Word of God was greatly enhanced by the invention of the printing press, as was the ability to become involved in comparative religious studies as the availability of books on other religions increased. And in the modern age, religion uses the communications systems of today to spread its message through television, radio, and computerized information systems. Scholars share information and search out obscure answers more quickly. Religious leaders and pilgrims travel with greater ease to the shrines of their faith. Whole new vistas of understanding open up in the face of a scientific community whose leading edge more and more approximates the mystic concepts of religions. Since science asks how, and religion asks why, they cannot be one, yet religion has the capacity to avail itself of modern technology like any other social institution.

And this is certainly not the only example that we can find. What about education? With the changes in technology leading to central heating, electric light, and the mass distribution of energy, the whole concept of education changes. As a result, the shift from the rural social structure to the urban one took place, necessitating a shift from the traditional, one-room schoolhouse of the last century to the mass education institutions of today. Indeed, education is ever in the throes of technological change. One hundred years ago, a person might spend five to seven years in the same schoolhouse, learning from the same teacher. Less than fifty years ago, education meant larger, more centralized schools, with more facilities shared by a larger number of students, and, hopefully, extended opportunities for gaining knowledge. Today, there is a shift away from the centralized approach, with college-by-television available to many students so that they need not even leave their homes in order to receive the education that they desire. The institution changes again.

And what of government? The political structure that a society chooses is also a social institution, whether it is the town meeting, diet, parliament, or congress. These are all political institutions, designed to carry out a single set of functions, that is, to govern (creating rules, supplying public goods and services, and maintaining order). Yet how does a structure change? Feudalism was killed by a population which settled as a result of the invention and implementation of the plow as a means of increasing productivity. Of course, feudalism took several hundred years to die once the death blow was struck; nevertheless, it was the lack of mobility on the part of the population that led to its decline. The rise of the middle class through trade and the expansion of knowledge created the right of the people to have a voice in their government. Although self-rule was not a new idea—indeed, the Greeks had a pure democracy long before the Romans ever thought of becoming an empire (I wonder how the Greek slaves felt about Greek democracy?)—the true rise of the middle class to political prominence came with the economic clout developed through its dominance of the technology of the age. Do we have democracy in the United States? Technically, the United States is a republic with a representative government rather than a direct government. People are elected to make laws rather than it being done by the population at large. Yet with changes in technology, such as television, computers, and rapid long-distance communications, the idea of every citizen voting on an issue before Congress is not as farfetched as it once was. What would be the effect on the political structure if everyone had a button on the television set to instantly vote on an issue?

And what of privacy? Privacy is a privilege that we take for granted in this country, yet it is strongly threatened by advances in technology. The ability of political and economic institutions to discover private information about individual citizens is awesome. There are satellites capable of focusing on a single individual on the ground. They can do so with such precision that the dial of a watch can be read from orbit. There is the capability, with the proper authorization, to screen telephone calls for certain key words and then record those conversations for later study, or to check on what people buy, what they owe, to whom it is owed, and whether they live beyond their means. These are capabilities considered by many to be threatening to the institution of privacy, a social structure long valued for its social value. Even the right to private property is altered by such simple technological devices as the photocopier and the tape recorder. How does one protect copyrights when it is so easy to gain illegal access to the fruits of one's labors?

Technology's Effect on Individual Freedom

Are we more free or less free by virtue of our technology? Apparently, the answer is yes and no. Depending on the use to which a technology is put, it can be either freeing or enslaving, or both at the same time. We are doomed to be dependent on technology as long as that technology is allowed to shape our world. The technology of building modern homes creates comfort, safety, beauty, security, and a host of other positive benefits that free us from the fears of our distant ancestors. Yet technology can enslave us to a life style that depends on this element being in our lives. How would it be if the houses were no longer there? How would we survive a cold winter? How would we protect ourselves from the elements? How would we maintain our privacy? We are dependent on housing in the forms that are available to maintain our life style.

A more serious example is our dependence on the automobile as a means of basic transportation. This dependency is one that has been brought home with shocking clarity in recent years as a result of shortages and rising fuel prices. When the fuel crisis of the 1970s began, many Americans were unaware of their dependency on the automobile to get them about. It was simply a fact of life, a technological device that was taken for granted in modern American society. But within weeks, motorists found themselves stranded in lines to buy gas, paying black market prices for fuel, wondering how they were going to cope with the crisis and how they could alter their life style. Seventy years of dependency on a reliable means of transportation had locked them into its use. In the long run there were solutions to that dependency. One could buy a foreign car that did not burn so much gas. One could buy a motorcycle or begin riding the local mass transit systems (a far more efficient form of transportation anyway), or take up bicycling or jogging. But in the short run, when the crisis first arose, all that could be done was to bite the bullet, dig a little deeper into the old wallet, and pray that the pump did not run dry before your turn.

As an experiment, consider the following. There is a simple device that is available to all of us. It is a technological miracle that has given us a tremendous amount of freedom in our lives, yet it has created a huge dependency as well. It is called the light bulb. To drive home just how freeing and how enslaving this simple device is, take note for the next few hours just how often you use one. Note each time you switch on a light in your home or office. Consider how often that light is on and that you have no control over how long it lasts. Consider street lights, the little light in your car that automatically goes on when you open the door, or the one in your refrigerator that always goes off when the door closes. How often and in how many ways does the lowly light bulb free you, yet make you dependent? Think about it.

Technology's Effect on Our Perception of Reality

An individual's perception of reality is created by the observations that are made by that person. If someone were forced to grow up in a room that was totally dark at all times and was never allowed to see any light, that person's perception of reality would be seriously distorted in that he or she would lack any content involving seeing. If another person were never allowed to experience kindness, that person would grow up believing that kindness in humanity is a myth. What we observe to be true, we believe to be true. In fact, we build our world in the image of those beliefs.

But what if the world we experience is altered by technology? What if we experience something new that does not fit into our contextual framework? This is the effect that new technology can have on a person's sense of reality, and it is no more than what was earlier discussed in terms of altering the paradigm with which we work. I can think of no more dramatic or elegant example of this phenomenon than the following, which is presented as the only example to be given.

Prior to the second half of the twentieth century, no known person had ever viewed the planet earth from a height of more than a few tens of miles. We were well aware of the planet's makeup, of what constituted its surface, and where the various planetary features could be found, but no one had ever seen the planet from afar. Then we entered the age of space exploration which brought with it, among other things, the first pictures of our own neighborhood. From the first moment that first camera took the first picture of that huge blue marble in a velvet void, the life of every person on the planet was changed. It was possible for the first time to see that we are all aboard the same ship traveling at some 250 miles per second through a sea of cosmic flotsam, and that what happens to one of us happens to all of us, particularly as concerning the planet itself. This is one ship that it is difficult to get off. It is this realization among an increasing number of people that holds the greatest hope for stability in the world of tomorrow.

Technology's Effect on Our Mutual Dependence

Buckminster Fuller once recommended ending the threat of nuclear holocaust by connecting the electrical grids of the United States, Canada, and the USSR to create a single, huge electrical system. He theorized that no one is crazy enough to blow up the other half of their own electrical grid.

Regardless of the merits of the idea, it illustrates an important aspect of technology and what life is in a technological world. Just as there is an increase in our dependence on technology, there is also the possibility of becoming more dependent on one another because of technological involvement. The United States is highly dependent on Middle Eastern oil producers for the supplies of crude oil needed to run our economy. Without oil, we would be hard pressed to maintain supplies of fuels, plastics, and a host of chemicals, just to mention a few items. In a similar manner, much of the world is dependent on the United States for food. Because of technological innovations in agriculture, less than 5 percent of our population is capable of feeding not only our own population but millions and millions of others. Nevertheless, if we are to have coffee, we must import it. If we are to have rare earths and exotic metals, much of the supply must come from elsewhere. We live in what is, as Marshall McLuhan has said, a global village. This is a *world* economy that we are involved with, and it is that involvement that makes us dependent on one another for what we need to survive.

Technology can both create and alleviate that dependency. In the absence of certain goods (such as, for instance, our dependence on the supply of natural rubber from Southeast Asia during World War II), technological innovation can create new substitutes. Likewise, a dependence on a certain technology may result in a dependence on a specific commodity, such as our insatiable thirst for oil stemming from combustion engine technology.

The key is to consider the consequences of a technology in terms of its tendency to increase or decrease mutual dependence of sociopolitico-economic groups and, from that, ascertain the probable outcome of a technology's introduction.

Technology's Greatest Effect

What sector of the population will be most seriously affected by a given technology and when? If the technology is one that can be generalized over the entire society, then it will probably have an impact on everyone, but if it has localized application, then what form of effect will exist and for whom?

Historically, any example that deals with a technology being localized for some reason would suffice to demonstrate the necessity of studying this issue. In the ancient world, about the time of the rise of the Mesopotamian city states such as Ur and Lagash in the fertile crescent, it was the technological innovation of agriculture that created cities where trade took place. City states arose around the farmlands of the Tigris and Euphrates river valleys, and because of the availability of water transportation (another example of technology), they were able to grow prosperous and powerful. Thus, the citizens of these areas were most greatly affected by the introduction of the agricultural technology, though tribal peoples from along the rivers were involved in the trade that resulted from the sedentary life style the agricultural people practiced.

A more contemporary example is the age of "king cotton" in the southern United States. With the cotton gin, cotton manufacture became cheap and profitable. By technologically solving the problem of how to comb the seeds from the cotton fiber rapidly and cheaply, the desirability of cotton as a crop rose dramatically, and there was a ready market for those goods in England, where the textile industry depended on cotton and wool for its livelihood. The people most directly affected by the technology were those in a position to take advantage of it, that is, the Southern states, where the climate was perfect for growing cotton and where the employment of slave labor was productive enough to be profitable. These conditions did not exist in the North, and, as a result, people living there were relatively untouched by either the cotton industry or the slave labor method of operation, since Northern industry depended on different inputs to create goods and services. Eventually, the schism in social systems represented by the localized nature of the cotton industry technology led to the Civil War and the end of an era. After years of political and economic dominance by an agricultural South, the nation was dominated by a more efficient and more productive North, where, it should be noted, there was a technological advantage in manufacturing industries because of the abundance of raw materials and transportation.

Whom a technology affects is as important as how. This can lead to moral questions as well. What of the wonder drugs that could be manufactured or developed but are not because the number of patients requiring them is too small to warrant the costs? What of the ability to save lives through expensive operations such as heart transplants or mechanical hearts? Who receives them and who does not? Who pays for them? Are there too few people affected to warrant continuing the technology? These are some of the questions that arise in considering this aspect of technological development.

CONCLUSION

Ten effects of technology have been presented. The reader can no doubt think of many more. Specific subjects not covered might include the effect of technology on health, on other technology, or on the security of the individual and the nation. These and other topics were considered for the set of subsets, and most were rejected as too narrow in scope, or able to be covered under one or more of the other headings. It is hoped that the reader can use the list as a systematic first step into the study of technology and society. From here, the book takes a slightly different turn, presenting an actual example of society in the face of technological growth. Included are two papers presented by students whose purpose was to investigate the effects particular technologies have had and continue to have on our lives.

THOUGHT AND PROCESS

1. Choose some simple technological device and speculate on how it has affected your life and the lives of people in your society. Pick something that seems insignificant, such as a pencil eraser or the common zipper. Itemize as many uses and applications for the technology as you can think of. Check your list against the list of social factors above and see how many you can find.

2. Choose an institution (such as education, religion, politics, the family, the extended family, or your own group of friends), and think about how technology has created that institution in its present form. Is this form different from those of previous times? Are you personally aware of any changes that have taken place in your experience of the institution? Which is better—the way the institution was, the way it is, or is it just different in the face of new conditions?

3. Ask a small child, a young adult, and an older adult about how they feel regarding (a) the computer, (b) space exploration, and (c) modern medical techniques. How do their responses differ? Who shows the highest degree of acceptance? Who the least? After reading this far in the book, how have your own feelings about modern technology changed, if at all?

4. This problem is "food for thought." One of the underlying scientific principles with which engineers, physicists, chemists, and others work every day is the circle of 365°. Trigonometric functions, esoteric topology, vector analysis, astronomical measurement, and industrial quality control are all areas of investigation that require some sense of understanding on the part of the investigator of how this circular system of measurement works. It is identical in action to the sweeping hands of a clock. Inasmuch as reading clocks is a symbolic-conceptual process by which we gain expertise in visualizing and manipulating the informational content of a circular coordinate system (that is, we learn to understand the circular coordinate system by an intimate familiarity with clocks as a symbol of that system), how do you think the growing predominance of digital clocks will affect future generations in their attempt to learn about angles and trigonometric functions?

5. Imagine the availability of a vehicle of about the same size and cost as the ordinary car that operates on the principle of levitation. It can operate at up to ten feet above the surface and is in all other ways comparable in performance and cost to the normal automobile. In other words, the only severe change in design

in the vehicle is the antigravity mode of suspension and propulsion. Speculate on how it will change life in our society, using the list of social elements given in this chapter.

World War II: Technologies and Social Implications

By
Eric Robinson
and
Russell Pennington*

World War II officially ended on the Battleship *Missouri* on September 2, 1945. It would be unrealistic, however, to assume that the effects of the war halted at the same time. In fact, we all live under the shadow of World War II, the greatest source of change in modern history. Virtually all technological developments have their roots in the war years. But even beyond this, humanity's philosophy of life has been drastically altered. All in all, the effects of World War II are much more dramatic and widespread than the postwar population can imagine.

Before we examine the effects of World War II on the modern world, let us first scrutinize the war itself and the ways in which this war was different from any other. The differences we are going to examine are not by any means inclusive; a discourse on every difference between World War II and any other war would run several thousand pages. What follows *does* represent the major developments through the war years. The major elements analyzed will be

1. Increased mechanization
2. Increased mobility
3. Increased role of science
4. Mobilization of the civilian sector

*Eric Robinson and Russell Pennington, "World War II: Technologies and Social Implications" (an unpublished paper presented at DeVry Institute of Technology, Atlanta, Georgia, April 17, 1984).

5. Improved communications
6. Total nature of war
7. Use of atomic weapons

INCREASED MECHANIZATION

World War II, unlike previous wars, was fought by human beings and machines, rather than simply human beings. This means that an army consisted as much of the factories at home producing the machines as of the soldiers in its ranks. The ability to produce effective weapons became the ultimate goal of the production lines. This led to a staggering growth in the use of sophisticated weaponry. Tanks and planes were employed as never before; the attack on Pearl Harbor was a far cry from the almost benign reconnaissance planes of World War I. The planes had far greater range and mobility, making bombing a favorite type of attack with a high chance for success. Tanks were no longer simply armored vehicles with thin metallic shells. Now they were as heavy as 36 tons, with 62-millimeter armor, and they could reach speeds of 25 miles per hour. Some tanks were armed with a massive 75-millimeter gun.[1] The numbers were staggering as well. In 1940, the United States produced 300 tanks, 6,100 planes, and ships with a total displacement of 52,600 tons. Production reached maximum levels of 29,500 tanks (1943), 96,300 planes (1944), and 3,176,800 tons of shipping (1944).[2] Some mechanization had no precedent—for the first time the world saw jets, rockets, amphibious units, and the atomic bomb. The increased mechanization was almost inconceivable to the people of the time and still boggles the imagination.

INCREASED MOBILITY

The increase in mechanization led directly to an increase in the mobility of both the military forces and the population in general. But the sharpest increase in mobility at the time was in the area of the ability of the military forces to move large quantities of soldiers, machines, and supplies. During World War II soldiers could be transported rapidly by rail, motor car, plane, or ship, and canned food could be brought to them. In earlier wars, only 1 percent of a given population could be

[1] Richard Collier, *The War in the Desert* (New York: Time-Life, Inc., 1977), p. 113.

[2] Edward Jablonski, *A Pictorial History of the World War II Years* (Garden City, N.Y.: Doubleday, 1977), p. 294.

mobilized at any one time. In World War II, there were mobilizations totaling 10 percent of some populations with as much as 2.5 percent at the front.[3] This meant that a much greater proportion of the people were directly involved in the actual fighting. The advances in technology created better means of communication, which enabled the armies to be virtually anywhere on the face of the earth and still be in close contact with their superiors at home, thereby giving the units more actual mobility. This increase in mobilization had many effects on the military and on the rest of the world, now and then.

INCREASED ROLE OF SCIENCE

The advances in weapons and mobility would not have been possible without a corresponding increase in the development of technology. These developments would have been impossible without science. As the war raged on, so did the race for the fastest, strongest weapons. Each army depended on its production lines and its scientists and engineers to keep it on the leading edge of technology. An example of need dictating development is the growth of the United States' tank force. The earliest models of the war were virtually useless against the German panzers—the guns could not penetrate the armor nor could the carriages outrun the heavier, faster German machines. Later models, such as the Sherman tank, surpassed the panzers in every way. The improvements occurred in a period of only four years. It can be said that World War II spawned a "blitzkrieg" of technological developments and scientific discoveries.[4]

MOBILIZATION OF THE CIVILIAN SECTOR

Since a greater proportion of the population of any country could be mobilized at any given time, a greater proportion of the population at home was required to be mobilized into military support services. For any army to be successful in its campaigns, the civilian sector must be dedicated to its support. This dedication is required during the war and in times of peace as well. The organization of the civilian sector by the military has become necessary as a preparation for war. This has resulted in the armed forces no longer being a self-contained unit apart from the general population. However, the military does not overwhelm

[3] John G. Burke and Marshal C. Eakin, eds., *Technology and Change* (San Francisco: Boyd and Fraser Publishing Co., 1979), p. 308.
[4] Collier, *War*, p. 113.

the civilian sector. For this reason it became necessary for the military to make use of propaganda to sustain the greatly needed morale of the population. This use of propaganda extends government control into virtually all aspects of the society, including the economy and public opinion. This in turn has led to the increase of the autocratic totalitarian state and the subsequent elimination of free speech and free economy in some countries. History has proved that a free-market system is less adequate than a military-based economy for the support of the war effort.[5] It is ironic that a war fought in the name of freedom has led to the reduction of freedom in many areas of the globe.

IMPROVED COMMUNICATIONS

As mentioned previously, the increased nationalism due to World War II increased the importance of propaganda. Propaganda is only effective when distributed to the masses—the greatest words of wisdom are wasted if no one is listening. It was therefore important to each government to ensure that every citizen have some way of "listening." This led, in part, to an explosion in the communications industry. Since armies were spread out further, better communications were also required to coordinate the movements of the armies. Improvements were made in radiography, radar, cryptography, and mass communications technology.

TOTAL NATURE OF WAR

The modern military technique called for the mobilization of the civilian sector. Since it is directly involved in the support of the military, it is obvious that the civilian sector would now become military targets and therefore be liable to attack from the enemy. The part of the population involved in the contribution to the war effort is not safe from attack. It should be apparent that the enemy has little or no way of distinguishing the difference between the military-supporting population and the nonmilitary population. This lack of distinction has led people to realize the total nature of war. The civilian population of World War II saw starvation, bombardment, confiscation of property, and terrorism, which have been considered since that time to be applicable against any enemy population as a whole. The advent of a possible attack against a nation's general population has caused the methods of war to become more ruthless and indiscriminate. Today the methods

[5] Burke and Eakin, *Technology*, pp. 308-10.

of war call for the use of terrorism, which is a direct result of the total nature of war.[6]

USE OF ATOMIC WEAPONS

The total nature of war is a startling and frightening concept. But the ante was upped just before the war ended. Humanity discovered an entirely new way to be inhumane to itself—the atomic bomb. The atomic bomb represents a whole new way to attack the enemy—mass destruction of property and people utilizing a single bomb. The atomic bomb had an additional impact in that it made defense essentially obsolete. The only defense became retaliation. This had the further effect of making the old war machine essentially fickle. A disquieting result of all this is that the smallest country can intimidate the biggest country if it has the "big one!" Yet this bomb was supposed to be a peace-keeping force. As Hans Bethe said, "If two opponents armed with hand grenades face each other in a six-by-nine-foot cellar room, how great is the temptation to throw first?"[7]

Unquestionably, World War II still affects us today. To simplify analysis, we will attempt to discuss aspects of society and technology in discrete parts. Naturally, all of the elements are actually interactive, and nothing affects one aspect while leaving the others untouched. The specific areas we will deal with are

1. Economics
2. Ecological-geographical
3. Sociopolitical
4. Discoveries
5. Ethics and morality
6. Psychological

ECONOMICS

Economics can be viewed in two ways: the immediate effects during the war and the long-range effects after the war. The war pulled our nation out of the deepest depression in its history and put the postwar

[6] *Ibid.*, p. 310.
[7] Karl Jaspers, *The Future of Mankind* (Chicago: University of Chicago Press, 1961), p. 59.

economy in a boom period that lasted for the next twenty years. The main change in the economy during the war came in the form of the billions of dollars in government contracts given to the corporations of United States. This, in turn, gave our nation the possibility for the staggering production that it demonstrated so well to the world. Another change during the war was the increased employment of women and the increased use of assembly-line techniques. The economic necessities of war benefited not only women, but Black Americans as well. By 1943 the number of skilled Black workers had doubled and the number of semiskilled workers rose even more steeply. Nearly two-thirds of the one million Blacks who took war jobs were women. "More women, especially married ones, worked for wages. The rate of the female work force participation rose 24%, peaking in 1944 with 19,370,000 working women. Women took jobs customarily allocated to men. They worked at steel mills, ship yards, airplane factories, and railroads."[8] The heavy industry jobs paid much more than the service jobs usually allocated to women. The war aided all aspects of business, but the large corporations received the largest benefit. The administrators of the federal war production efforts were usually the heads of large corporations. These administrators usually gave their own companies the government contracts whenever possible. Thirty-three corporations received one-half of $75 billion in contracts between June 1940 and September 1944. The war centered the power of the corporate administrations, and as a result corporate profits rose from $6.4 billion in 1940 to $10.8 billion in 1944. The war's boom was so intense that even the most severe strike wave in the history of the United States couldn't stop it. The strikes of 1945 to 1946 involved 4.5 million workers in 6,000 separate strikes. During this period corporations won the right to pass on wage concessions to the consumer.

In Europe, the economy was on the verge of total collapse. Despite U.S. aid to Western Europe and Japan tataling $14 billion beginning in 1945, by 1947 Europe was economically on its knees. The loans were given mainly to Great Britain and France. Great Britain received $3.7 billion. France received $1.4 billion. These amounts were merely a drop in the bucket; neither France nor Great Britain were even close to their prewar production levels. To make matters worse, the most severe winter in many years hit Europe in 1946 to 1947. Production fell 50 percent in Great Britain alone. The United States realized that something had to be done, so the Marshall Plan was born. This plan called for long-range economic aid to be given to the war-ravaged countries of Europe. The plan called for the continent of Europe to be

[8] Melvyn Dubofsky and Athan Theoharis, *Imperial Democracy: The United States Since 1945* (Englewood Cliffs, N.J.: Prentice-Hall, 1983), pp. 5-6.

treated as a whole, not as individual countries. The Soviet Union objected to the Marshall Plan for that reason. This disagreement contributed to the cold war between the United States and the USSR for the next twenty years. If any one thing was affected by the war, it was the economic condition of the world, now and then. Even today and for many years to come, we find our economic situation stemming from decisions initiated as a result of World War II, such as the Marshall Plan.[9]

ECOLOGICAL-GEOGRAPHICAL

World War II changed the face of the earth. Also, anyone who bought a map prior to the war tried valiantly to get a refund, until it was discovered that it was a collector's item. Germany was divided into four sectors immediately following the war. This division has been a source of tension ever since it was made. In general, the shape of many European countries changed. Beyond the way the land was represented on maps, there were changes in the land itself. Europe, Japan, and Russia were all war-torn. When they rebuilt, they neglected to replace farm land. Throughout the world, there was a marked decline in the amount of farm land, and a still greater decline in the number of farmers. Both of these conditions stemmed from the increased role of factories and automation, and the resultant decline in human labor. Finally, humanity created for itself another environmental concern—that of radioactivity. By the end of the war, people around the world began to realize that as destructive as World War II had been, the next war could be the last.

SOCIOPOLITICAL

The startling worldwide realization that a single country could conceivably destroy the world led to a greater interdependence of nations. That is, the Monroe Doctrine was declared null and void. Isolationism would no longer work in a constantly shrinking world. Following World War II, the United Nations, NATO, and the Warsaw Pact were chartered. Reluctance to join was based on the fact that these would be merely entangling alliances, not unlike those that were the root cause of World War I. Yet the peoples of the world, as a whole, needed some symbol of unification following the traumatic war; the UN represented a compromise between the two factions. As one would well expect, the

[9] *Ibid.*, p. 22.

tragic war led to an aversion to war; this unwillingness for foreign involvement at least partially accounted for the "fall of China." This led
almost directly to the infamous "Red Scare." After the brutal leader-
ship of the Axis powers in World War II, it was easy for the American
public to group all aggression together as a "plot." Certainly, the space
race has its roots well-entrenched in World War II. The Red Scare and
technology gained from the war provided both the motivation and the
means for space exploration. Finally, the horrors of war, and especially
the Holocaust, led to a new emphasis on the fundamental rights of all
humans. A part of the United Nations' charter discusses the issue of
human rights, and that all humans have certain inalienable rights. This is
where the civil rights and women's movements have their roots as well.

DISCOVERIES

World War II put the world in a mad scramble for technology. The
echoes of this scramble are still being heard today. During the war,
there were many new discoveries. They included jet engines, rockets,
synthetic fuel, plasma, penicillin, advanced pesticides, and nuclear
power.

The Germans were losing the war, but they did not quit. They
worked furiously at coming up with some new device that would save
them from the advancing Allies. One of the things that they developed
was the jet engine. Obviously, the jet has changed our world. The jet
has made global travel convenient and economical. It has made war machines that can fly three to four times the speed of sound; without the
jet our world would be totally different. This is not to say that without
World War II the jet would not have been developed; the war merely
sped up the technological research required for that development.

The rocket was developed almost exactly in the same fashion as
the jet. Hitler's scientists designed the rocket as a vengeance weapon.
The V-1 "buzz bomb" and the V-2 rockets were the weapons that would
periodically obliterate one or possibly two city blocks in London. The
rockets were not mass produced, so the Germans could not really
exploit the power of this technology. The rocket technology proved
invaluable in the years following the war. President J.F. Kennedy employed Hitler's chief rocket scientist in the beginning of the space race.
The roots of our space program and the roots of our nuclear missiles
are in the early German rocket research conducted by Wernher von
Braun and his colleagues.

The war saw many shortages, but the most drastic shortage as far
as the military was concerned was a shortage of fuel for internal combustion engines. In fact, the Germans lost the Battle of the Bulge as a

result of a severe fuel shortage. Scientists on both sides were working on the development of a synthetic fuel substitute. Toward the end of the war, one was discovered, but the problem was that the cost proved to be too high for practicality. This synthetic fuel never saw use, yet in the future, when the world will have used its supplies of crude oil, the synthetic fuel research of this period can be used. Today research is in progress to improve on the fuel formulas of this era.

Any war is going to have casualties. World War II was no different. The doctors were trying to find better ways to deal with the wounded. There were several medical developments during the war. The top two discoveries were the use of blood plasma and penicillin. Both were great medical innovations, possibly the greatest that we have seen in this century. The medical breakthroughs could have come at any rate, but undoubtedly the war sped up the research.

Part of the research toward the war effort was in the area of generally useful chemicals for many applications, military and otherwise. There were many chemicals developed, but the most useful from this area were in the field of pesticides. Some of the pesticides developed were later proven to be environmentally unsafe, yet the research proved to be useful in many areas, such as in the development of plastics and rubber.

The greatest development of the war, or the most infamous, depending on one's perspective, was nuclear power. The immediate effects are well known—the annihilation of thousands of Japanese people. The long-range effects are not yet fully understood as the controversy over nuclear energy rages on. The only thing that can or should be said is that the issue affects every single human being today and in the future.

None of the developments mentioned above was due singularly to the war, but all of them were altered by the war. During the war, the development of all of them was sped up, and the research on them was intensified.

ETHICS AND MORALITY

World War II raised some serious questions in the minds of many people. In this portion of the paper we deal with the ethical and moral impact of the war. The implications in this area are literally infinite in number, as each person has his or her own questions or considerations concerning the impact of the war. Some of the questions that we feel are applicable to this paper deal with the technological implications in the realm of ethics and morality.

Where are science and technology leading us? This is a very important consideration. Do we know where science is leading us? After the war, many people came to the realization that the technology around them was virtually in control of their lives. Humankind and its machines were inseparable. Never before in all of the history of the human race had there been a time like that following World War II. Suddenly people discovered that the whole globe could be destroyed in a matter of days. This was, to say the least, a difficult concept to grasp. The "explosion" of technological research during the war demonstrated to the world's peoples that technology was growing at a rate much faster than their ability to comprehend it. The conclusion that many people reached was that they were not in control of their own existence. This was one way of looking at the concept, but more than likely the problem was that, for the first time, humankind was totally in control of its destiny.

The human race was startled to realize that technology, its creation, was leading the world, not the other way around. Of course, this is only one perspective; the other would say that we are the masters of technology. The latter is much easier to deal with in the rational human mind. We made it, therefore we can control it. Since the first concept is more difficult, we will deal with it here. Do we want to go where technology is leading us? Many people wonder where the world is headed and if the scientists and engineers are suited to determine the world's destiny. There has always been a resistance to change, but the changes in the period of and immediately following the war were of a magnitude never before experienced. There are no simple answers. There are no answers at all in the usual sense! These issues are simply things that the person of our time must consider.

If humanity was to decide that it needed or wanted to control technology, could it? For the Allied powers to win World War II, they pushed for massive scientific research. They did not question its worth to society in general. The only thing that they cared about was, Would a particular research end the war a day sooner? The lasting effects of the technology were not considered; it was simply a necessity. This "open season" on research led the world to a point of no return. The more technology human beings have, the more they crave. This has become a vicious circle. Should or shouldn't we try to halt technology is not the question; the question is can we? We could or would not voluntarily go back to a more primitive way of life. The right or wrong of the issue is irrelevant. The fact remains that human beings are destined to continue in their technological development.

One of the more ordinary questions resulting from the war was that of human rights. This idea is not new. People were forced to look at the way they thought about other people and how other people viewed them. People began to wonder if their race could be the Jew of the next Nazi Germany. Thus the idea of human rights was renewed.

Today we believe that people—all people—have certain rights that are not uncommon to any race or nation. The interest in the concept of human rights sparked the beginnings of the modern civil rights movement in the United States and has altered the entire world's thought patterns.

PSYCHOLOGICAL

It comes as no surprise, then, that as human beings have changed their thought patterns about the world in which they live, they have suffered psychological repercussions. The realization that the world could be destroyed any time has led to the idea of instant gratification. The idea that since the world can go at any time, the thought, Why shouldn't I do just what I please? has led to the moral decline of the population. Typically, all people follow a natural progression. As children, we seek instant gratification. As we gain maturity, we begin to see that we cannot always have what we want; sometimes we have to wait for it. But the shadow of the "big one" has made us less willing to wait. As learning patience is a major part of maturity, the society's development has been interrupted by the harsh realities of the world around us. Many people cannot handle the lack of control that they feel around them. They seek escapes in drugs, alcohol, or any other thing to suppress their fear of what they cannot control. Others react differently. Feeling that they have no control, they assume the attitude that they do not care, that they are not concerned with what happens to them or with the world. Or they believe that their input is so valueless that there is no point in giving it. Voting records indicate that since World War II, there has been a marked increase in the "undecided" or "what difference does it make?" response to voter surveys. Yet another reaction, and probably a more healthy one, is the realization that what happens on one end of the globe can affect people on the other end. This makes the population hungry for information as to what is happening. They have a fear of the unknown. They demand more information, faster and more accurate. They want to see it from the source. This has led to the explosion of what is commonly referred to as the "Information Age." Naturally, there are other aspects influencing the processing and distribution of information, but its roots are at least partially in the postwar reaction to the war with Germany.

It would be understating the issue to merely contend that World War II was a dominant force in every person's life forty years ago. We believe a more accurate statement is that World War II could be thought of as a motivating factor driving the general population to a period of increased productivity and innovation. It could well be contended that the impetus was a little overzealous, but this point must be considered by the readers. It is our purpose merely to serve as messenger, bringing up the issues for each person to consider on his or her own. For, in the end, each person will form his or her own evaluation of the implications of World War II, and the way it has affected that individual's life.

BIBLIOGRAPHY

Bauer, Yehuda, *The Holocaust in Historical Perspective.* Seattle, Wash.: University of Washington Press, 1978.

Burke, John G., and Marshal C. Eakin, eds., *Technology and Change.* San Francisco: Boyd and Fraser Publishing Co., 1979.

Collier, Richard, *The War in the Desert.* New York: Time-Life, Inc., 1977.

Dubofsky, Melvyn, and Athan Theoharis, *Imperial Democracy: The United States Since 1945.* Englewood Cliffs, N.J.: Prentice-Hall, 1983.

Hoehling, Allan, *Home Front U.S.A.* New York: Thomas Y. Crowell, 1966.

Jablonski, Edward, *A Pictorial History of the World War II Years.* Garden City, N.Y.: Doubleday, 1977.

Jaspers, Karl, *The Future of Mankind.* Chicago: University of Chicago Press, 1961.

Solar Energy and Its Social Consequences

By
Kevin Bagwell
and
Wayne Ergle*

ABSTRACT

Solar energy is and has always been an abundant source of alternative energy. Furthermore, the sun will continue to shine whether we choose to take advantage of the energy or not. Several astounding innovations have already taken place in the solar energy industry. These include the development of a solar-powered automobile, a solar-powered generating plant, solar heating for the house, and solar energy in space.

The social and economic issues of solar energy are also quite important. The use of solar energy would result in a cleaner and safer earth. Unfortunately, cost is still a major setback in the field of solar energy. In order to maintain the rate of advancement in the solar industry, the government and society must choose to provide the researchers with the funds needed to develop new technology that will cut the costs of solar energy. Solar energy will be a primary alternative source of power in the future. The natural resources of the earth are being consumed at an astounding rate. It is up to the present society to give the green light to further the research in the solar energy field before all of the natural resources are depleted.

*Kevin Bagwell and Wayne Ergle, "Solar Energy and Its Social Consequences" (an unpublished paper presented at DeVry Institute of Technology, Atlanta, Georgia, May 2, 1984).

INTRODUCTION

This paper contains a discussion on solar energy and its social consequences on contemporary society. As is the case with all alternative sources of energy, scientists have just begun to realize the true potential of using the sun as an efficient source of power.

In order to properly discuss these social consequences, several factors are introduced throughout this paper. Some of these factors include:

1. The technical innovations developed in the solar energy industry within recent years. These innovations will directly affect the way the society comes to depend on solar energy both today and in the future.
2. The social issues of solar energy based upon contemporary society. These issues will include the government's views on solar energy, the public's reaction to those views on solar energy, and the advantages and disadvantages of a solar-reliant society.
3. The economic impact of using solar energy for an alternative source of energy. This section will include a discussion on the effects of solar energy on the physical environment, the rate of development of a society based on solar energy, and how solar energy will provide for the advancement of economic structures.

DEFINITION OF SOLAR ENERGY

Solar energy has always been a readily abundant source of alternative energy. The concept of solar energy applies to every life-sustaining organism known to humankind. However, what is solar energy? Is it the visible light spectrum seen by the human eye? Or is it the invisible radiation resulting from the solar activity of the sun? By deriving a general definition of both solar and energy, the term *solar energy* can be thought of as the "natural energy vigorously exerted by the action of the sun's light and/or heat."[1]

THESIS STATEMENT

Humanity has depended on solar energy since the creation of the earth. The primitive human beings possibly discovered the use of solar energy by laying on a sunlit rock for warmth. Or perhaps they found that their

[1] David B. Guralnik, ed., *Webster's New Collegiate Dictionary* (Cleveland, Ohio: Collins Publishers, 1975).

clothes would dry much faster when exposed to the sunlight. The ancient civilizations increased the use of solar energy by using the sun to bake the clay bricks used for building materials. However, to write a paper discussing all of the aspects of solar energy since the beginning of time would be a monumental task. Therefore, this paper will be restricted to the issues of solar energy in contemporary society. This will allow the discussion to focus on how solar energy technology has affected the present generation. Through the technical innovations of solar energy within recent years, contemporary society has become more conscious of the reality of solar energy as a primary source of alternative energy.

TECHNICAL INNOVATIONS

Some of the more useful innovations of the solar energy industry are geared toward contemporary society. These innovations include solar heating, solar-powered airplanes and automobiles, solar-powered generating plants, and solar technology for space equipment. The following paragraphs discuss these innovations in detail and describe the feasibility of each innovation from a societal viewpoint.

Solar heating actually consists of two types, active and passive. An active solar heating system actually uses solar collectors to heat a substance (usually water or air) that is circulated through the house via heat registers. When using an active system, a storage area is needed to store the unused heat. Usually the storage area can retain the heat for three to five days of cloudy weather. The active system operates by using solar collectors (commonly located on the roof or adjacent ground) to focus the sunlight on a system of pipes contained inside the collector. The pipes contain the heating substance, which is then transferred into the storage area until heat is needed. When the need for heat arises, the substance is pumped out of the area and circulated through the house in a manner similar to a normal HVAC system. The substance is then pumped back through the solar collector and transferred to the storage area once again, until heat is needed.

One of the more obvious setbacks to the active system is that if cloudy conditions exceed more than five days, a backup source of energy (such as propane or natural gas) is necessary to provide the desired heat. Another setback is the initial cost of an active system. A totally dependent solar house can be extremely expensive to build. Fortunately, federal and local governments of many states offer grants or tax credits to offset the initial investment in the system.

The passive solar heating system retains the same basic concepts of the active system without the need for expensive equipment. Basically, the passive system uses large windows located on the east and west exteriors of the house to allow the sun to warm walls and floors constructed of masonry or slate. This process provides heat for both daytime and nighttime hours. At night, insulated shutters are closed over the windows allowing the radiant heat contained within the walls to be released, thereby heating the rooms. This concept is very practical in cold weather areas that experience a great deal of sunshine.

Most people think of a solar house as a house consisting solely of apparatus used by the sun. However, in most cases, solar heating should be combined with other equipment to provide additional efficiency and savings. Heat pumps, wood burning stoves, and well-insulated walls and floors all help to provide a truly energy-efficient home.

Israel is one of many countries to experiment with the development of a solar-powered automobile. The engineering department of Tel Aviv University has developed a solar-powered car named the *Citicar*. The Citicar was designed by a group of students under the direction of Professor Arye Braunstein.[2] This vehicle is capable of traveling at speeds of up to forty miles per hour and has a maximum range of fifty miles for each charge. The car operates on a two-step DC system (24/48 volt) utilizing two solar panels mounted on the roof and hood of the car. These panels consist of 342 solar cells and provide a peak power of 400 watts and a charge to the battery of 48 volts. The entire weight of the car including batteries and solar panels is approximately 1,320 pounds. Improvements are planned to provide an additional ten miles per hour to the top speed and to double the effective range of the car per charge. Since Israel is located in one of the sunniest regions of the world, the Citicar should prove to be beneficial to the society. Unfortunately, no estimated cost for this vehicle was discovered during the research of this topic.

In 1981, the Solar Challenger (a solar-powered airplane) successfully completed a five-hour-and-twenty-three-minute flight over the English Channel. This was the first time a solar-powered plane had completed a flight using only the sun as its source of energy. The Solar Challenger was designed and constructed by Paul MacCready of Pasadena, California.[3] The Challenger generated its power from 16,128 solar cells located on its wings and horizontal stabilizer. These cells powered two electric motors connected to the propellers of the plane. Using tough, lightweight plastics (Kevlar, Mylar, and Lucite), the Challenger weighs only 217 pounds unoccupied. Unfortunately, the cost of the

[2] "Israel's Solar-Powered Car," *Mother Earth News*, vol. 65 (September–October 1980), p. 120.

[3] "Icarus Would Have Loved It," *Time Magazine*, vol. 118 (July 20, 1982), p. 45.

flight exceeded $725,000; flying by solar power commercially could be a long time off into the future.

Solar One, located near Barstow, California, is one of the first ten-megawatt solar-powered generating plants in the world. The plant consists of "over 1,800 heliostats, sprawled in a fan-shaped array covering 70 acres, focused on the central receiver, a 45-foot-tall metal cylinder atop a 300-foot tower."[4] The plant was designed and built by the McDonald Douglas Corporation and is operated by the California Edison Company under a government contract. The cost of the plant totaled over $141 million. Solar One doesn't turn solar energy into electricity. Instead, the plant uses solar thermal technology, which uses the heliostats (concentrated mirrors) to focus the sun onto the central receiver containing a vapor. This vapor is heated to 700°F, which is fed through turbine generators. These generators then produce the actual electricity that is sent to the power grids. This system is now undergoing a five-year test program. However, with the favorable results that have already been recorded, California Edison is planning to construct a larger 100-megawatt plant to be in operation by 1988. Like any new innovation, there have been minor setbacks. Some of these setbacks include nonfunctional heliostats, unexpected thermal gradients, and buckling of the panels on the central receiver. A storage area (consisting of a bed of rocks or high-temperature oil) can retain heat for a period of four hours after the system goes down to continue the production of power. An automatic control system is being designed so that the entire plant can be operated with a crew of four employees. Within time, engineers should solve the problems to date, making the idea of a solar-powered generating plant a costworthy investment.

Using solar energy for space vehicles is one of the most efficient concepts available. Most of today's satellites contain solar cells or panels that provide the energy needed to operate the onboard equipment. Both the original Skylab and the present Space Shuttle contain solar panels that generate the power used to operate the craft. With the successful repair of the satellite on the last shuttle mission, it is now apparent that solar energy has become even more feasible for the satellites of the future. Now for a fraction of the original cost, companies will be able to recover their nonoperational satellites and implement repairs on these satellites in space, saving the companies millions of dollars.

Scientists are now working on an idea known as the solar power satellite (SPS). This concept is a satellite that would contain either a photovoltaic array or a thermal power plant. The satellite would trans-

<hr/>

[4] "Solar One; Sun, to Heat to Electricity," *Popular Science*, vol. 221 (October 1982), p. 114.

mit the electrical power generated to antenna receivers located on the earth. These receivers would then transfer the power to power grids, which would distribute the power to consumers. There are a number of positive arguments for such a system. The system would eliminate the hydrocarbon combustible products, radioactivity, and second-layer thermal pollution that is now occurring on the earth. Since the satellite would be in space, it would have an unobstructed sun and would also be immune to factors on the earth such as earthquakes, severe weather, and so forth. However, for every positive argument there appear to be two negative arguments. The earth-bound antenna receivers will have to be massive in size. The system will theoretically introduce water vapor and microwaves into the upper atmosphere. The largest argument against the system is the astronomical cost. A prototype satellite is estimated to cost $1.3 trillion under the most favorable conditions. Perhaps in the near future scientists will find a way to reduce the cost of the system. The general attitude of the government at this time is that the system is just too costly for such a gamble.

SOCIAL ISSUES

There are numerous issues concerning the society that arise when discussing solar energy. Perhaps the primary issue is, What are the social consequences of accepting solar energy as a source of power? By placing the cost of solar energy aside for a moment, we can observe several beneficial aspects of solar energy technology. First of all, solar energy is available to every geographic region of the world. It does not require any special resource found only in a particular region. Naturally, some regions of the world are sunnier than others. However, with the implementation of the solar power satellite system, the actual amount of sunshine that a country receives becomes irrelevant. There is a great possibility that solar energy will allow a country to become self-sufficient in energy resources. If this becomes true, the resource-poor countries will not have to rely on the resource-rich countries to provide them with power at an exorbitant price. This concept can also be related to the individual level. It is not theoretically impossible than an individual will own and operate his or her own power plant based on personal needs. This concept has already become evident in solar heating. This would also allow the individual to become more self-sufficient as far as energy needs are concerned.

The U.S. government's views toward solar energy were at one time favorable. In the United States, there are still grants and tax credits available in some states that encourage the implementation of solar equipment. A solar bank was proposed by the Jimmy Carter administration providing subsidizing and the lowering of interest rates for solar energy construction. On the proclaimed "Sun Day" in 1978, Carter called for the "dawning of the second solar age."[5] However, in the Reagan administration, officials have already begun to cut the Energy Department's budget for solar energy research to half its former level. In 1982, they unsuccessfully tried to eliminate the tax credit available for investing in the solar energy industry. Fortunately, even with the resistance by the Reagan administration, homeowners and investors still feel that the solar energy industry is worthy of their investments.

ECONOMIC IMPACTS

Solar energy is one of the cleanest and most abundant sources of energy available to humankind. The sun will continue to shine regardless of whether we choose to take advantage of the energy or not. One of the few drawbacks of using solar energy is that, theoretically, the amount of water vapor and microwaves will increase within the upper atmosphere. However, the advantages of solar energy far outnumber the disadvantages. First of all, solar energy is clean. There are no by-products or hazardous wastes encountered when using the sun. Perhaps best of all, the physical environment of the land is not harmed in any way. The most fundamental change in the ecological base would be that a cleaner and safer countryside would result. Without the combustion engines that pollute the atmosphere, much of the air pollution would cease to exist. Since the demand for natural resources such as petroleum and coal would decrease, the physical environment would not continue to suffer from strip mining or oil spills that occur today.

It is extremely likely that solar energy would provide countries with a new set of goods and services. Designing, developing, and maintaining solar equipment would come into demand. Products such as solar energy equipment for the house would become popular. However, other industries would suffer. Any industry in the field of producing internal combustion products would find the demand for its services severely decreased. Many other industries would have to adapt to the new energy source. Automobile manufacturers, power companies, and the oil industry would have to change their products to supply the new needs of the society. The oil industry has already begun to prepare for the future. Many of the oil companies have invested in various areas of the solar technology industry.

Unfortunately, these economic impacts are strictly theoretical in

[5] "A Possibility, Not a Novelty," *Time Magazine*, vol. 114 (July 2, 1979), p. 27.

the present society. It seems probable that countries will continue to deplete and deface the natural environment until all of the natural sources of energy are exhausted. Only then, it appears, will countries seriously consider the alternative of solar energy. Let us hope that the natural environment can withstand the abuse until this decision is finally made. Until new technology finds a way to dramatically decrease the cost of using solar energy, a totally solar-reliant society is far from reality.

PREDICTIONS FOR THE FUTURE

The future of the solar energy industry can only be based on personal opinion due to the unknown future advances of technology. If the government decides that it is a worthy investment to support solar energy technology, then the expansion of the field is unlimited. On the other hand, if the government continues to cut back on the funds provided for solar energy research, the field will probably experience limited advancement until all other sources of energy are near exhaustion. Therefore, it is up to the government and the society as a whole to determine if solar energy research is technically feasible, cost-efficient, and worthy of future research. In a strictly hypothetical situation, based on the supposition that advanced technology and cost efficiency have been discovered, the following ideas are not entirely impossible.

It is possible that the solar power satellite system could provide the earth with enough solar power to make petroleum-powered generating plants obsolete. However, these plants could be converted to accept the thermal power of the SPS system, since the concepts are basically the same, allowing the plants to continue operation. I believe that the majority of housing will become solar-reliant. In Shenandoah, Georgia, entire subdivisions are being constructed using both active and passive solar energy systems. The recreation center in Shenandoah uses the sun to completely operate its HVAC system. This also includes the freezing of the ice for the skating rink. It is possible that entire communities will own and operate their own solar power stations. To carry the concept further, individual homeowners who live in isolated parts of the region could implement an antenna system (similar to a communications system) that would allow the homeowner to receive electrical power directly from the satellite.

There appears to be little doubt that many of the items used by consumers will use solar energy for power. There are already numerous items on the market, including pocket calculators and small novelty items, that do so. It does not seem impractical to expand this idea to larger household items such as the lawnmower or any other motorized machinery. Solar-powered automobiles will become more evident in the future. These vehicles will provide the society with a safer, more efficient, and cleaner form of transportation.

CONCLUSION

Solar energy is an alternative source of energy that seriously needs to be considered. There will come a time in the not-so-distant future when the physical resources of the earth will either be too expensive to recover or completely exhausted. Solar power allows for a cleaner environment and reduces the chances of catastrophe that might occur with other sources of energy such as nuclear energy. However, research in the solar energy field will be limited unless the U.S. government decides to give its support to the idea that solar energy is a solution to the problem. With the technological advances that occur every day, there appears to be little doubt that research in the solar energy field will dramatically lower the costs of solar equipment. This paper has discussed some of the technical innovations resulting from the extensive research in the solar energy industry. The automobile, solar heating, and space technology are all actually in use today. With these results already established, one can only imagine what the future holds in the industry.

The social consequences of solar energy are primarily beneficial to the society. The use of solar energy would increase and strengthen an individual's independence. Implementation of solar products would decrease the costs of the power and provide a cleaner and more efficient way to operate the home. Quite possibly the best result of solar energy will be a cleaner and healthier earth.

Therefore, the society must make a relatively straightforward decision: Is it worth the extra research expense in the field of solar energy to increase the chances of a cleaner environment? Or is it more beneficial to the society to continue the exhausting consumption of natural resources, contend with the increasing amount of pollution, and hope that the resources will not be exhausted until after we have left the earth and let the future generations contend with the problems we have bequeathed to them?

BIBLIOGRAPHY

"A Possibility, Not a Novelty," *Time Magazine*, vol. 114 (July 2, 1979), p. 27.

Guralnik, David B., ed., *Webster's New Collegiate Dictionary*. Cleveland, Ohio: Collins Publishers, 1975.

Hayes, Denis, "Environmental Benefits of a Solar World," *Vital Speeches of the Day*, vol. 46 (March 1, 1980), pp. 306–10.

"Icarus Would Have Loved It," *Time Magazine*, vol. 118 (July 20, 1982), p. 45.

"Israel's Solar-Powered Car," *Mother Earth News*, vol. 65 (September–October 1980), p. 120.

Lee, Al, "Passive/Active Solar," *Popular Science*, vol. 219 (December 1981), p. 123.

"No Profit in Politics," *Environment*, vol. 24 (March 1982), p. 25.

"Oil Industry Buys a Place in the Sun," *Science Digest*, vol. 221 (August 26, 1983), p. 839.

"Solar One; Sun, to Heat to Electricity," *Popular Science*, vol. 221 (October 1982), p. 114.

"Solar Power Satellite: A Plea for Rationality," *Science*, vol. 203 (February 23, 1979), p. 709.

"Solar Plane Soars," *Popular Mechanics*, vol. 155 (April 1981), p. 127.

"Tax Breaks Help Fuel Solar Power," *Business Week* (April 26, 1982), p. 36.

PART III

Conclusion

CHAPTER 9

Prelude

Most books begin with a prelude rather than end with one. Yet due to the nature of this work, the prelude comes at the end, as a summary of what has come before, and as an introduction leading to what will come in the future. Since where we have been is no more than a preface to what will be, this book ends at the beginning of the next step, and the next step is the reader's.

Unlike descriptive texts, the purpose of this book has been to acquaint the reader only with enough primary information to start the search for knowledge. Like technology itself, that search is always new, always just beginning, and always punctuated with the imagination and hopes of the investigator. Technology is not a "thing" per se. It never has been. It is a process, a natural consequence of who and what we are. It is a way of surviving, growing, and thriving in the face of the ever-changing environmental conditions that challenge our ingenuity to the limit. If it were not so, then we would surely have succumbed to the pitfalls of this planet long ago.

In the first chapters of the book, we dealt with the nature of technology, searching for a definitive understanding of what it was we were studying, and analyzing the "nature of the beast," in an attempt to give ourselves direction. We considered the artificial nature of technology, the fact that it consists of constructs that are unnatural manipulations of the physical laws of the universe in which we operate, and that, in spite of this unnatural content of device and artifact, the *context* in which the artifacts were held was completely natural.

We create technology by natural design. We manipulate the laws of nature in accordance with our understanding of those laws, and bend them (they cannot really be broken) to fit our needs. Technology is a mirror, an extension of humanity, created in its image to increase the species' capacity to function. The wheel is nothing more than the foot taken to its logical conclusion, then pushed further by compounding the technology of the wheel with that of the piston, a mechanical arm. Likewise, steam or gasoline power is an extension of solar energy in a transformed state to provide the desired results—cheap, efficient power and mobility for the human race. The camera and the lens are no more than specialized eyes, yet very special inorganic eyes that can see further than the eyes of the sharpest eagle, and detect movement more adroitly than the wisest cat. And ears are extended through such devices as microphones, amplifiers, and sonar equipment. And the bird's secret is harnessed in the airplane, the jet liner, and an Atlas missile in flight. And the dolphin is emulated through submarines, the turtle through tanks and mobile homes, and the ant and the termite through cavernous structures lying both above and below ground through which we pass in constant streams to carry out the collective and cooperative duties of a modern society in full operation.

And yet in each of these cases, the emulation is only superficial, humanity adapting the lessons of nature to be learned from other living creatures and from ourselves to push a principle artificially through technology as far as it will take us. We are builders. We are dreamers. We are the ultimate simulator in nature, borrowing from everywhere to combine what nature does through selection and then purposely structuring the pieces into a cohesive whole for our own use.

Thus we build our technology. And in so doing, we speed up the process of evolution many thousandfold, doing more in fifty years with the concept of flight than nature was able to do through natural selection in 70 million. With such an increase in speed comes hazards, and with the hazards the need for a check to balance the growth with caution. And homeostasis provides us with that check. By resisting change, and by limiting our acceptance of the new and tempering the fever for growth with the fear of the unknown, humanity has been able to select those technological innovations that create value without too much inherent cost. The balance is maintained, at least for the present, through the conflict between the desire to extend ourselves and the fear of rushing into unknown realms. And so it has been for thousands of years.

Along with this dual concept of innovation and homeostasis, we viewed the nature of the creativity that feeds the process, the condition of insatiable thirst for knowledge that seems to permeate every aspect of human life. We considered curiosity. We considered the changing paradigms. We considered the revolutionary nature of an animal ruled not by genetic coding and uncontrolled conditioning, but by a freewill choice to restructure the normal into the new. Like any good scientific investigators, we studied the "how" of the process, and then, in a contextual shift, we considered motivation, the "why" of technology as well.

Economic impact is a single segment of a total social pattern that includes technology and is both created by and creates the physical mechanisms by which we are able to maintain our lives. The why and the how of technology are initially economic in nature, expressing our desire to increase the overall welfare of the society through the production and distribution of goods. And as we saw, those goods are the result of production utilizing *scarce* resources, resources that are not easily replaced and which increase in the effort required to obtain them. It is the technology that allows us not only to expand our productive capacity, filling the society with an ever-widening variety of consumption items, but also to continue producing in the face of dwindling supplies. Efficiency—the capacity to produce with a minimum of expense, effort, or time—is the why of technology, as is consumption. Economically, everything that we have in the way of goods and services is directly attributable to some aspect of technology.

In the second part of the book, methods of study were explored, including a look at the usefulness of cause and effect in constructive thinking and the use of models in developing an understanding of the nature of the processes taking place in the complex modern world. The systems view was presented as one logical way of organizing information and structuring our understanding of technological processes, emphasizing the interrelationships that exist among the various facets of society as it reacts to technology and the consequences of our own ingenuity.

If we are to understand the place of technology in our world, we must have some means, no matter how unreliable, of being predictive, of describing the expected consequences of our actions in an attempt to deal with the obvious pitfalls of technological progress before they catch us unawares, knowing full well that no matter how effective we are at prescient descriptions of future consequences, the unseen and unconsidered results still need to be handled as they arise. That we are not perfectly predictive in our attempts at prescience is an indication of our humanness. So is our felt need to be as predictive as possible, to weigh the odds in our favor by considering as many possible future worlds as we can, and by preparing for those contingencies that can be imagined and are deemed likely enough and urgent enough to require our attention now, before they become a major problem in our lives.

Which brings us to the here and now, the so-called prelude. Why the prelude? Why not the epilogue or the afterword? Why end a study with a beginning? Because.

Or more precisely, "be-cause." To this point the nature of a technological society has been defined, and methods for study have been discussed. But the actual study of the subject is yet to begin. That is no accident. It was never the intention of this book to tackle the immense content of technology and society, to pour over the essence of weighty tomes, searching for some significant body of truth to pass on to the reader, as if by the magic of the summary of 10,000 years of history, any book could capture the meaning and nature of social change. To do so would require a detailed understanding of mechanics, physics, biology, botany, medicine, engineering, anthropology, history, religion, economics, philosophy, archaeology, and a hundred other disciplines. And in the end, the only presentation that could be made is the opinion of the author, a contextual understanding of technology's impact on society, restricted to the paradigms embraced by some group of investigators. So I repeat: "Be-cause."

Be the cause of your own experience of the subject. Involve yourself in the content of technological change and of humanity's use of technology. Find your own answers to your own questions, and *be* the *cause* of the future. Not only is this chapter a prelude, indeed, the entire book is a prelude. It is the beginning of the reader's own experience of what it is like to live in and understand technological societies. It is hoped that through personal investigation, some sense of the immensity of the subject and some understanding of the beauty and fascination inherent in the discipline will begin to implant themselves in the mind of the reader. Tomorrow is always before us, and we must be always preparing for its arrival. To know what is now true is to be left behind. As John Naisbitt has indicated, the proper frame for success today is the future. What will you be doing five years from now? Where will you be working? Where will you live? What will you think, and what will you wear? These are very real questions, and for the vast majority of the population they are questions without adequate answers. What will your reaction be to the technology of the future? How will you behave? How will you be restricted by the technology that controls your life? How will you control your life through that technology? Will you even exist five years from now? How do you know?

Obviously, you do not. But an understanding of the nature of the future is the best bet that any of us has to help ensure that future and our place in it. Thinking that emphasizes where we are going is necessary. Just as a jet pilot must think five or ten minutes in the future in order to keep up with his or her craft, so must we think five, ten, or even twenty years in the future to keep up with our world. And the progress of the world is one thing that will not wait for any person, should he or she fall behind.

By looking at the past and finding the patterns that repeat themselves, we can adequately operate in the present. By understanding the conditions under which we are now living we can prepare ourselves for the near future. And by studying the technology of today and the developments of tomorrow, we can cause the future to happen. We can form it and guide it, leading our lives into channels of success and away from the unseen dangers of dead-end social behavior that could leave us stranded as the culture passes us by.

And what are the possibilities of that future? We stand on the edge of a great adventure, one that will take us within ourselves, within our planet, and outward toward the stars. We have the capacity to explore the reaches of near space and the ability to reach and learn from the deepest recesses of the ocean floor. Underway right now is research that can give us the ability to produce needed chemicals through biological machinery created through genetic splicing and biological manipulation. We are on the verge of curing a dozen killer diseases, including significant gains in the battle against cancer. We have the ability to feed the people of the world, and in the face of that ability, millions of people suffer from starvation and drought. We can reattach a severed finger or arm with some hope of the patient regaining partial or total use of the limb. We can replace hearts and kidneys, eyes and vital tissue. We can make deaf children hear and watch them experience their own laughter for the first time. We have the capacity to reach out and mine the asteroids, to look across the universe to the beginning of time, to span an entire continent in a few hours, or to calculate the probabilities of success in a horse race while sitting at a terminal in our own home.

We have other capacities as well. We can kill our fellow human beings more efficiently than ever before. We can strike with colorless, odorless gas, or viruses against which there is no defense. We can incinerate entire populations or create a winter that would outlast the last human being. We can turn the planet over to the insects and the mosses, and we can do it all in no more than a single twenty-four hour day. And we are improving our capacity to do this as well.

How we choose to spend the rest of our lives depends to a large degree on how we perceive that future. A Chinese sage, famous for his study of war and the art of warfare, said some 2,500 years ago that we should know our enemy and know ourselves. To do so is to enter into battle without fear. Not knowing one or the other is disastrous in the extreme. The future is our battlefield, and our technology is what we must know. To know it and understand its potential is to give ourselves the power to extend the development of the human race to heights not yet dreamed of. We are the future. What we decide to do in terms of technology is what will create that future for us all. It is a responsibility and a legacy that we will all leave to those who follow. To ignore the consequences of future actions until they are upon us is to guarantee disaster, for ourselves and for all of our fellow travelers on this journey. Neglect is tantamount to imposing an unknown sentence on us all.

The important issue is whether or not to "be cause." It is a choice of watching society progress, being involved in society's progression, or causing that progress to whatever degree we choose. For this reason, and with hope for the future of us all, the prelude ends. To the reader is left the opportunity to fill in the first word of the first line of the first paragraph of your own . . .

CHAPTER ONE
TOMORROW AND BEYOND

APPENDIX

Topics for Discussion and Analysis

The following is a partial list of suggested topics for study in the field of social issues in technology:

Chemical warfare—an alternative to the bomb?

Euthanasia—the result of improved health care?

Sex discrimination and technological change

Age discrimination—the result of improved health care?

Technology and the world theologies (the relationship between religion and changes in technology)

Pollution and the Malthusian proposition

Technology and changing ethical patterns

Radio and alteration of the culture

Television and alteration of the culture

The plow and medieval society

Economic consequences of medical technology in the twentieth century

The postindustrial society (information-based economies)

Technology and music—fundamental changes

Edison—historical impact of his inventions, life, approach to research, and so forth

Social implications of space exploration and colonization

Social implications of orbital manufacturing

Science fiction as a predictive device—society's protection against "future shock"?

Creativity and freedom: Does culture create technology or vice versa?

Housing, technology, and culture

Technology and its effects on art and industrial design

Railroads and the building of the American society

History of glass (cultural effects)

Robotics

Time, technology, and society

The changing computer

The use of bioengineering in medicine

Computers as educators: Do they do the job?

Computers and time compression: Do they change our perception of reality?

Technology and the changing face of World War I

Technology and the changing face of World War II

The laser—a solution looking for a problem

Chinese technological advances—the industrial revolution that never was

Refrigeration

Waterways and technology—the transportation impetus

Ship building and the advancement of social systems

Hypothetical technology in a Utopian society

The printing press and its logical consequences

The computer and systems thinking

The age of steam and its social consequences

Nuclear energy and social consequences

Solar power and social consequences

Geothermal power and social consequences

The importance of freedom of expression in the technological society

Air transport and the modern society

The development of the electrical automobile

The discovery and cultural impetus of microwave technology

Artificial intelligence and the nature of creativity

Hygrophonics—a new science?

Hydroponics—the farm of the future?

Nuclear waste—its disposal and its threat

Acid rain

Nikola Tesla: Genius or charlatan?

The social impact of Charles Steinmetz

The role of lighter-than-air craft in commerce: past, present, and future

Medical technology in the twentieth century

Leonardo da Vinci—his life and influence

The microscope and development of the microcosm in human thinking

The telescope and changing paradigms

Renaissance science and the Church

Architecture and technology

The Roman legion and engineering

The future of cryogenics

Countercultures as homeostatic reactions to technological change

Photography and changing perceptions of reality

The library of the future—form, content, and accessibility

Technology as a function of sexuality

Plastic and the changing face of society

Technology and the changing nature of language

Dam-building techniques and the spread of culture

Technology as a reflection of available resources

Glossary

Adaptability A characteristic an organism is said to have if it can change its structure to meet the needs of a changing environment, or alter its method of operation to fit new circumstances in that environment. The more adaptability an organism has, the greater its capacity to survive in changing environments, and the wider the range of environments that it can inhabit.

Adjustment In psychology, the act of modifying, changing, or adapting individual behavior so that it conforms to generally acceptable societal norms.

Analog model A model that behaves in some manner similar to the reality that it is designed to represent.

Anxiety In psychology, a vague feeling of apprehension and hope for the future resulting in tension and stemming from some external stimuli or internal, unresolved conflict.

A priori knowledge Knowledge thought to exist in the mind exclusive of knowledge. Knowledge from a valid law to a particular incident.

Artifact Any object made by a human being that is particularly designed to be used in some way that increases efficiency and reduces hardship for the user.

Beatnik Any member of a counterculture existing primarily in the 1950s, characterized by subculturally distinct speech and clothing, and professing a dislike of accepted social behavior patterns.

Biological growth curve The sinusoidal curve form representing the growth and decline of living organisms through time.

Bohemian Any member of a counterculture existing primarily in the 1920s and early 1930s, exhibiting behavior considered at the time to be irresponsible. The phenomenon is thought by some to have been the result of an inability on the part of the members of the subcultural group to cope with sociological and technological changes beginning with World War I.

Capital investment   The investing of savings from individuals in the means of production in order to increase the capital base of the economy and thus create economic growth and an expansion of available goods for final consumption. It is a method of foregoing present consumption to ensure higher levels of consumption later.

Capitalism   An economic system distinguished by self-interest, consumer sovereignty, capitalization, competition, private property, and utilization of the market system to solve the problems of production and distribution within the economy.

Cause and effect   A relationship between actions and events in which the events are the result of the actions.

Competition   The concept in capitalism that both consumers and producers must vie with each other for the right to scarce resources and finished goods, and that through this process, a free market system is able to properly produce and distribute goods and services through the society.

Conditioning   In psychology, the process of acquiring or learning certain modes of behavior through reward and punishment, repetition, or other learning methodology.

Consumer sovereignty   The concept that the consumer has the final authority in deciding what will be produced, who will produce it, how it will be produced, and for whom it will be produced in a free market economy.

Content   The physical reality of something. Content is the real-world constituent makeup of a given object or situation, and is measurable and discernible in physical terms.

Context   The view held by an individual concerning the meaning, condition, or characteristics of the content of some situation or physical object. It is the individual "idea" that a person attaches to a physical reality by virtue or his or her own opinions, beliefs, experiences, and so forth.

Counterculture   Any subculture whose behavior patterns, mores, and customs are counter to those of the general culture within which they exist. It is usually a reaction to some change in the societal structure with which the countercultural members are unable or unwilling to embrace.

Creativity   The condition of causing to come into being as an original and non-naturally occurring phenomenon. To produce a unique and novel device, idea, or work.

Custom-made model   A model that is created for a specific need, designed to perform specific functions as a model of the reality that it is designed to represent.

Descriptive model   A model designed to describe characteristics or conditions of some reality that it is designed to represent.

Deterministic model   A model designed to offer a definitive answer to questions concerning some reality that it is designed to represent. This type of model is generally, though not exclusively, mathematical in form.

Early majority   That part of the majority of the buying public that purchases a new product that has survived the test of the innovative minority early in its introductory stage. The early majority purchases during the phase of the product life cycle represented by sales rates that increase at an increasing rate.

Economics   The study of how people choose to use scarce resources for the production of goods and services and how they choose to distribute those goods and services to the general population for their consumption.

Economic trade-off   The concept that states that because of the scarcity of resources, the choice of any given economic activity reduces the opportunity to carry on alternative economic activities by an amount equal to the resources flowing into the chosen activity. That is, for each choice to produce and consume a given product, we trade the opportunity to do other things with the same resources, labor, and capital inputs.

Economic welfare   The overall economic well-being of a sociopoliticoeconomic system. It is a nonquantitative factor, though it can be described in quantitative terms, dealing with the quality of life experienced by individuals in the system.

Economies of scale   The principle that, as production rises, the cost of production per unit declines due to increasing efficiency with which inputs are used. That is, as production rises, costs rise at a decreasing rate.

Efficiency   The quality of producing or working with a minimum of effort, time, and energy.

Entrepreneur   An individual who is aware of an opportunity to make a profit by buying low and selling high in the market. These are the risk-takers of the society who instigate economic behavior and production of goods and services, seeking a profit.

Evolution   The continuous adaptation of species to changing environments through selection, mutation, and hybridization.

External evolution   The adaptation of a species to changing environments through means external to genetic change, as in humanity's use of technology to create adaptive change external to the human body.

Feedback   Information returned to an instigator of change that reflects the results of that change, allowing the originator to correct for undesired results.

Frustration   In psychology, the condition of thwarted drives, in which attempts to move toward the satisfaction of some drive is denied, or attempts to escape from some negative drive are denied. The result of frustration is often anxiety and the need to adjust.

Game   Any recreation or activity incorporating specific rules which requires the participants to compete in some way, either against other players or against themselves in attempting to achieve some specific goal.

Hierarchy of needs   A motivational theory developed by Abraham Maslow, in which human motivation is seen as action designed to satisfy certain needs. The needs, from most basic to most advanced, are physiological needs, safety needs, social needs, self-esteem needs, and self-actualization needs. According to the theory, the individual tends to forsake satisfying higher-order needs in order to protect the fulfillment of threatened lower-order needs, thus creating the hierarchy of influence.

Hippie   Any member of a counterculture existing mainly in the mid-1960s to mid-1970s, whose members exhibited characteristics of pacifism and a denial of generally accepted societal goals. Some believe this countercultural movement to be the result of collective revulsion of public policy surrounding the Vietnam War, particularly as affected by the extensive television coverage of events in Southeast Asia during this period.

Holistic thinking   A nonlinear approach to thinking in which events and environments are viewed as a whole rather than concentrating on discrete parts. Philosophically Eastern, this type of thinking deals with the concept of everything being a part of the whole rather than discrete individual physical realities unto themselves.

Homeostasis   Resistance to change. The tendency of humans to resist change due to fear, a desire to reduce work, or an inability to understand the change in question.

Iconic model   A model designed to resemble a physical reality, though not to behave in an analogous manner. These models look like the physical reality that they are designed to represent

Industrial concentration   The degree to which productive capacity is concentrated in some small number of firms within a given industry. It is usually measured by the concentration of the top four or eight firms involved, with a monopoly of one company with all productive capacity representing maximum concentration and pure competition representing minimum concentration.

Innate ability   Inborn traits, such as aptitudes or physical characteristics, giving rise to specific dominant capabilities in the individual. These abilities are genetic in source.

Innovation   The introduction of something new, or recreation of some established technology into different or novel forms.

Innovators   Those who innovate. In marketing, those members of the society willing to try new products because of their newness. The least homeostatic of the consuming population.

Input   That which is introduced into a system to be processed for purposes of producing some output. Input can be in the form of raw materials, labor, capital, information, or any other phenomenon that carries some form of content.

Instinct   An inborn pattern of activity or tendency to some specific form of behavior common to all members of a given species.

Late majority   That sector of the consuming public willing to purchase a product only after it has become an accepted item. This portion of the population purchases goods near the maturity point of a product's life cycle, when sales are increasing at a decreasing rate. This is the most homeostatic sector of the consuming public.

Law of diminishing returns   After a given point, additional units of input will result in progressively smaller quantities of additional output, with total output first increasing at a decreasing rate and finally decreasing absolutely. That is, past a certain maximum level of input efficiency, each additional unit of resource input to the productive process will be less efficient than the unit prior to it.

Law of downward-sloping demand   There is an inverse relationship between the price of a product and the amount of a product that will be demanded in the marketplace, all other things being equal.

Law of supply   In general, there is a direct relationship between the price of a good and the amount of a good that is offered in the market by producers.

Linear thinking   A mode of thinking that views reality as a set of discrete physical objects and events, whereby the rule of cause and effect is followed. This is step-by-step thinking incorporated into Western logic.

Luddite   Any of a group of workers in England between 1811 and 1816 who roamed the countryside destroying machinery and technological innovations in the belief that this machinery was responsible for destroying jobs. The movement receives its name from one Ned Lud, an eighteenth-century worker who first exhibited the destructive tendency due to frustration caused by an inability to deal with automatic equipment in the textile mill in which he was employed.

Malthusian curve   The sinusoidal curve form representing Malthus's contention that, due to the manner in which food supplies and population grow, the human race was doomed to cease growing and possibly decline as a result of starvation, disease, and war.

Malthusian proposition   The theory presented by Thomas Malthus that indicated that since population grows geometrically and food supplies grow arithmetically, eventually population would outstrip the food supplies, and starvation would ensue.

Malthus, Thomas   The first political economist. An eighteenth-century economist who investigated the relationship between economics and population. It is after Thomas Malthus and David Ricardo, a contemporary, that economics receives its title of the "dismal science."

Market equilibrium   A condition in a free market that is said to exist if buyers and sellers agree on the quantity of goods to be bought and sold and on the price at which they will be bought and sold, and agree in addition to go on buying and selling that quantity at that price.

Maslow, Abraham   Twentieth-century psychologist responsible for developing the hierarchy of needs to explain human motivation and behavior.

Maslow hierarchy of needs   (*see* Hierarchy of needs)

Mathematical model   A symbolic manipulative representation of reality designed to describe relationships among certain factors of the reality that it is designed to represent. The method uses numerical representations to describe the reality in question.

Memory trace   The connective "highway" created among brain cells as a given memory is created. A memory trace is reinforced whenever the memory is activated, and weakened when not reinforced through time.

Model   A copy of a physical structure or a concept that is designed to demonstrate certain characteristics of that physical structure or concept in accordance with the purposes of the modeler.

Motivation   In psychology, that which prompts or drives a person to exhibit a given pattern of behavior.

Natural selection   A process occurring in nature by which only those organisms with traits particularly favorable to survival are able to survive in the competitive environment. They are therefore "selected" for survivability through the process.

Negative externalities   Costs incurred by third parties external to the economic decisions made by buyers and sellers of some product or service, and incurred because of the market interactions of these groups.

Oligopoly   An economic market structure characterized by high industrial concentration, difficulty in entering and leaving the market, and the sale of a uniform product or product group. This form of market appears to be a natural result of high production efficiency inherent in technological societies experiencing economies of scale.

Optimizing model   A decision model designed to present the best solution possible to the modeler, that is, how to "optimize" results in line with certain predetermined criteria. It is often, though not necessarily, mathematical in nature.

Output   The result of the production process. In systems theory, that which results from the processing of inputs. It is the goal that the system has as its purpose for existing.

Paradigm   A belief system that limits action by creating outside parameters within which one is allowed to operate. These are descriptions of reality based on observation that are used to explain that reality, excluding alternative explanations until replaced with some broader or different understanding of the phenomena in question.

Physiological needs   In the Maslow hierarchy, those basic needs that enable a person to maintain life, including warmth, food, water, shelter, and so forth. They are the lowest-level needs in the hierarchy.

Primary effect   The immediate and obvious effect of some action or event, particularly in relation to simulation building and systems analysis.

Private property   The characteristic of capitalism that holds inviolable the right of individuals to hold the means of production as personal property and the right to do with that property as they wish.

Probabilistic model   A nondeterministic mathematical model designed to predict the most likely outcome of some set of activities. It is designed to create a "best guess" of the probable behavior of the reality that the model is designed to approximate.

Process   In systems analysis, that factor of the basic systems format that refers to actions taken by the system to convert inputs to acceptable and sought outputs. It is the method by which the goal of the system is achieved.

Production possibility curve   A graphic representation of economic trade-off, indicating the opportunity costs of producing a given product in terms of the amount of another product or product group that must be foregone in order to do so.

Product life cycle   The normal progress of a product in the market through four stages of life—introduction, growth, maturity, and decline. The cycle approximates a sinusoidal curve.

Psychophysiological restructuring   The process of relearning in which old patterns of habitual behavior are replaced with new ones. The restructuring occurs of necessity if homeostasis is to be overcome due to the established brain cell connections, analogous to memory trace patterns, causing a resistance to change.

Qualitative model   A model designed to encode inexact concepts numerically through the use of statistics to produce probabilistic descriptions of the reality that the model is designed to represent.

Quantitative model   A model designed to produce discrete values for characteristics of the reality that the model represents through the process of manipulating formulas.

Ready-made models   General models designed to produce answers to questions that are repetitive in application and often encountered.

Safety needs   In Maslow's hierarchy of needs, the second level of needs in which the individual is concerned with protection of self and avoidance of danger.

Satisfice   To satisfy a need to the point that other needs offer more compelling motivation for fulfillment, at which point the original need is abandoned in favor of the more important ones. In Maslow's hierarchy, the act of relieving the pressure of a need and moving on to the next one, even though the other needs are not totally satisfied.

Scientific method   A systematic method of investigating real-world phenomena designed to ensure consistency of approach and maintenance of the integrity of evidence and findings. It is the rational method for explaining real-world observations.

Secondary effects   In systems analysis and simulation theory, those effects stemming from an action that are not immediately obvious or are caused by the primary effects becoming a causal factor.

Self-actualization   In Maslow's hierarchy of needs, the highest level in the need hierarchy—the need to express who and what one is. It is the ultimate level of need fulfillment, at which a person is considered to be balanced and whole.

Self-esteem   In the Maslowian hierarchy of needs, the need to feel good about one's self and what one does, the need for self-respect and a feeling of self-worth. It is directly below self-actualization in the hierarchy and is considered to be one of the developmental needs that humans possess.

Self-interest   The capitalistic principle that states that each person in the economy acts in such a way as to maximize his or her own interests, and that it is because of this that the free market system produces the most goods at the lowest possible price and of the highest possible quality, making those goods available to the largest number of people.

Shortage   In a market structure, when demand exceeds supply at the prevailing price of the market. In a purely competitive market system, this results in a rise in price toward equilibrium.

Simulation   A model that copies the behavior of some aspect or aspects of reality. Its purpose is to describe and predict behavior in real-world situations within the confines of the parameters of the simulation model itself.

Social needs   In the Maslowian hierarchy, the central level of needs in the five-tier structure incorporating the belonging needs of individuals, including group membership, the need to succor, and the need to nurture. They are both physical needs and developmental needs, thus representing the boundary between the more basic and the higher needs in the structure.

Specialization   In biology, differentiation in order to adapt to a changing environment or some special set of circumstances.

Subsystem   A system that is part of a larger system and therefore represents a discrete element in that larger system. It represents a subclassification of activities in an overall systems design.

Surplus   A case of supply exceeding objective demand in a market system. When surplus exists in a purely competitive free market system, the usual result is a drop in price back toward equilibrium.

Survival of the fittest   The concept that those life forms best fitted (suited) to prevailing environmental conditions tend to survive, whereas those related organisms that are less fitted to survival tend to become extinct. It is an expression of the natural selection process of evolutionary theory.

Survival trait   Any trait that an organism possesses that tends to improve that organism's chances for survival. A trait that adds to survivability.

System   A collection of physical and nonphysical interactive parts all having connection through a mutual purpose of achieving some goal.

Systems analysis   The method of analyzing a system by breaking it up into its constituent parts and relationships in an effort to understand its structure and its ways of functioning, and to predict its behavior under given sets of conditions.

Systems approach   A method of studying a real-world phenomenon that centers on constructing a model of the reality in a systems mode, that is, describing a real-world phenomenon as a system. The advantage is to create an orderly structure and predictive ability through an understanding of the nature and goals of the phenomenon. The shortcoming is that it tends to create a paradigm that disallows possible explanations outside the orderly systemic form utilized to study the phenomenon in question.

Technological specialization   The process of narrowing one's scope, either as an individual or as a society, toward specific technological specialties, creating a high degree of division of labor; a separation of the worker from the process of work; and the accompanying psychosociological problems such as boredom, psychosomatic illness, dissatisfaction with job, apathy, reduced productivity, and a general decrease in psychological health.

Technologize   The act of creating technology. Although the resulting artifact is artificial by definition and not a natural structure, the act of creating is natural to humans and is a survival mechanism innate to the organism.

Technology   That whole collection of ways in which the members of a society provide themselves with the material tools and goods of their society. The collection of artifacts and concepts used to create an advanced sociopolitico-economic structure.

Verbal models   Models that are designed to convert thoughts and concepts into language, to establish relationships and restrictions of real-world systems, and then to organize them.

Bibliography

CHAPTER 1

Ardrey, Robert, *African Genesis: A Personal Investigation into Animal Origins and Nature of Man.* New York: Dell Pub. Co., Inc., 1961.

Birdsall, Derek, and Carlo M. Cipolla, *The Technology of Man: A Visual History.* London: Wildwood House Limited, 1979.

Cooke, Jean, Ann Kramer, and Theodore Rowland-Entwistle, *History's Time Line: A 40,000 Year Chronology of Civilization.* London: Grisewood & Dempsey, Ltd., 1981.

Dobzhansky, Theodosius, "Man and Natural Selection," in *Classics of Western Thought,* vol. 4, ed. Donald S. Gochberg. New York: Harcourt Brace Jovanovich, 1980.

Fincher, Jack, *Human Intelligence.* New York: Putnam's, 1976.

Jastrow, Robert, *The Enchanted Loom: Mind in the Universe.* New York: Simon & Schuster, 1981.

Matson, Floyd, *The Broken Image: Man, Science, and Society.* Garden City, N.J.: Doubleday, 1966.

CHAPTER 2

Alland, Alexander, *The Human Imperative.* New York: Columbia University Press, 1972.

Childe, V. Gordon, *Man Makes Himself.* New York: The New American Library, 1961.

Freud, Sigmund, *The Future of an Illusion*, trans. W. D. Robson-Scott. Garden City, N.J.: Doubleday, 1964.

Hays, Samuel P., *The Response to Industrialism: 1885–1914*. Chicago: University of Chicago Press, 1957.

Schaffer, Lawrence Frederic, and Edward Joseph Shoben, Jr., *The Psychology of Adjustment: A Dynamic and Experimental Approach to Personality and Mental Hygiene*. Boston: Houghton Mifflin Co., 1956.

CHAPTER 3

Bradley, Alan, *Your Memory: A User's Guide*. New York: Macmillan, 1982.

Bronowski, Jacob, *The Origins of Knowledge and Imagination*. New Haven, Conn.: Yale University Press, 1978.

Hassan, Ihab, *The Right Promethean Fire: Imagination, Science, and Cultural Change*. Chicago: University of Illinois Press, 1980.

Kuhn, Thomas S., *The Structure of Scientific Revolutions* (2nd ed). Chicago: University of Chicago Press, 1970.

Mote, Frederick, *Intellectual Foundations of China*. New York: Knopf, 1971.

Sahal, Devendra, *Patterns of Technological Innovation*. Reading, Mass.: Addison-Wesley, 1981.

Thomson, George, *The Inspiration of Science*. Garden City, N.J.: Doubleday, 1968.

Zukav, Gary, *The Dancing Wu Li Masters*. New York: Morrow, 1979.

CHAPTER 4

Allingham, Michael, *General Equilibrium*. New York: John Wiley, 1975.

Miller, Roger LeRoy, *Economics Today* (4th ed.). New York: Harper & Row, Pub., 1982.

Rasmussen, David W., and Charles T. Haworth, *Elements of Economics*. Chicago: Science Research Associates, Inc., 1984.

Smith, Adam, *An Inquiry into the Nature and Causes of the Wealth of Nations*, ed. Edwin Canaan. New York: Random House, 1937.

CHAPTER 5

Bowden, Elbert V., *Economic Evolution*. Cincinnati, Ohio: South-Western Publishing Co., 1981.

Stein, Philip, *Graphical Analysis: Understanding Graphs and Curves in Technology*. New York: Hayden Book Company, Inc., 1964.

CHAPTER 6

Bedemeier, Harry C., and Richard M. Stephenson, *The Analysis of Social Systems*. New York: Holt, Rinehart & Winston, 1962.

Roberts, Nancy, et al., *Introduction to Computer Simulation: The System Dynamic Approach*. Reading, Mass.: Addison-Wesley, 1983.

Rubenstein, Moshe F., and Kenneth R. Pfeiffer, *Concepts in Problem Solving*. Englewood Cliffs, N.J.: Prentice-Hall, 1980.

Springer, Clifford H., Robert E. Herlihy, and Robert I. Beggs, *Advanced Methods and Models*, Vol. 2, Math for Management Series. Homewood, Ill.: Richard D. Irwin, 1965.

CHAPTER 7

Bertalanfy, Ludwig von, *General Systems Theory: Foundations, Development, Applications*. New York: George Braziller, Inc., 1968.

Lynch, Robert E., and Thomas B. Swanzey, eds., *Examples of Science: An Anthology for College Composition*. Englewood Cliffs, N.J.: Prentice-Hall, 1981.

CHAPTER 8

Bel Geddes, Norman, *Horizons*. New York: Dover, 1977.

Galbraith, John Kenneth, *The New Industrial State*. Boston: Houghton Mifflin Co., 1967.

Graham, Loren R., *Between Science and Values*. New York: Columbia University Press, 1981.

Mumford, Lewis, *The City in History*. New York: Harcourt Brace Jovanovich, Inc., 1961.

Richter, Maurice N., Jr., *Technology and Social Complexity*. Albany, N.Y.: State University of New York Press, 1982.

Toffler, Alvin, *The Third Wave*. New York: Morrow, 1980.

CHAPTER 9

Bronowski, Jacob, *A Sense of the Future*, ed. Piero E. Ariotte. Cambridge, Mass.: The MIT Press, 1980.

Burke, James, *Connections*. Boston: Little, Brown, 1978.

Dyson, Freeman, *Disturbing the Universe*. New York: Harper & Row, Pub., 1979.

Hall, Edward T., *The Hidden Dimension*. Garden City, N.J.: Doubleday, 1969.

Morris, Desmond, *The Human Zoo*. New York: McGraw-Hill, 1969.

Naisbitt, John, *Megatrends: Ten Directions Transforming Our Lives*. New York: Warner Books, Inc., 1982.

O'Neill, Gerard K., *2081: A Hopeful View of the Human Future*. New York: Simon & Schuster, 1981.

Toffler, Alvin, *Future Shock*. New York: Random House, 1970.